Freshwater Fishes

Vertebrate Palaeobiology and Palaeoenvironments Set

coordinated by
Eric Buffetaut

Freshwater Fishes

250 Million Years of Evolutionary History

Lionel Cavin

ELSEVIER

First published 2017 in Great Britain and the United States by ISTE Press Ltd and Elsevier Ltd

ISTE Press Ltd
27-37 St George's Road
London SW19 4EU
UK

www.iste.co.uk

Elsevier Ltd
The Boulevard, Langford Lane
Kidlington, Oxford, OX5 1GB
UK

www.elsevier.com

Notices

Knowledge and best practice in this field are constantly changing. As new research and experience broaden our understanding, changes in research methods, professional practices, or medical treatment may become necessary.

Practitioners and researchers must always rely on their own experience and knowledge in evaluating and using any information, methods, compounds, or experiments described herein. In using such information or methods they should be mindful of their own safety and the safety of others, including parties for whom they have a professional responsibility.

To the fullest extent of the law, neither the Publisher nor the authors, contributors, or editors, assume any liability for any injury and/or damage to persons or property as a matter of products liability, negligence or otherwise, or from any use or operation of any methods, products, instructions, or ideas contained in the material herein.

For information on all our publications visit our website at http://store.elsevier.com/

British Library Cataloguing-in-Publication Data
A CIP record for this book is available from the British Library
Library of Congress Cataloging in Publication Data
A catalog record for this book is available from the Library of Congress
ISBN 978-1-78548-138-3

Printed and bound in the UK and US

Contents

Foreword

Since the pioneering work of Louis Agassiz at the beginning of the 19th century, fossil fishes have been the subject of many scientific works. Their evolutionary history has been reconstructed with a fair degree of accuracy, although welcome new discoveries are constantly emerging, providing new information and challenging our ideas. However, contrary to other groups like mammals or dinosaurs, this history has not often been discussed in a comprehensive summary that would allow us to situate it within the broader framework of the geography and environments of our planet, and one might well wonder why, because fishes are excellent material for such studies.

Lionel Cavin's book addresses this gap by presenting the reader with a summary of the evolutionary history of freshwater fishes over the last 250 million years since the start of the Triassic period. This long interval of time was characterized by major events in the evolution of these animals including the appearance of new groups and the succession of various faunal assemblages, as well as major events in the history of the Earth and living organisms including the break up of Pangaea, significant climate change, sea level variation and large-scale biological crises. It was during this period that the modern ichthyological world developed, an event that is no less biologically significant than the evolutionary radiations of mammals or birds. With such particular habitats and lifestyles determining the possibility of geographical dispersal, the history of freshwater fishes during the Mesozoic and Cenozoic eras generally reflects the major changes to the geography and environment over time. With his vast experience in paleoichthyology, Lionel Cavin retraces a history that is no less eventful than that of other groups of organisms that have attracted more attention. This makes this book quite a unique addition to the genre, of interest to fossil fishes experts as well as readers who are more generally interested in major evolutionary phenomena.

Eric BUFFETAUT
Director of Research Emeritus at the CNRS
Ecole Normale Supérieure de Paris

Introduction

Today, freshwater fishes account for one quarter of the total number of vertebrate species and half of the number of fish species. However, the volume of the environment that they occupy represents only one ten-thousandth of the total volume of water on Earth. How long has this massive diversity existed and how did it come to be? The increased fragmentation of continental environments, the isolation of river systems and latitudinal and altitudinal climatic variations that characterize these areas are just a few of the causes of this diversity. These features have varied over time, but it is likely that the fragmentation of freshwater ecosystems was already significant by the Mesozoic and Cenozoic eras. Among the important factors for the evolution of freshwater fishes, three have varied consistently since the beginning of the Mesozoic era: the global paleogeography structured by tectonic movements, climate and sea levels. In a very general sense, the first of these factors is associated with the fragmentation of Pangaea, the foremost tectonic event of the last 250 million years. At the start of the Triassic period, the supercontinent was an almost complete land mass that gradually broke apart to form the continental arrangement we know today. During that time, continental fragmentation was scarcely more significant than it is today. Only at the end of the Cretaceous, when India was still an island and Africa had not yet made contact with Eurasia, may our planet have displayed a more fragmented geography than it has now. This period was also a fruitful time for tetrapods, especially dinosaurs. The two other factors, climate and sea level, are probably more significant than the general positions of tectonic plates for understanding the evolutionary history of freshwater fishes because they influence the capacity for life and dispersal in shorter time frames. However, they are more difficult to characterize precisely, especially in the ancient periods that concern us here.

In this book, we will examine, on a global scale, the groups of freshwater fishes that have a fossil record dating from the Mesozoic and Cenozoic eras, a period covering the last 250 million years of the Earth's history. We will focus on the paleogeographical and paleoenvironmental contexts that surround the fossils and attempt to situate these discoveries within lengthy evolutionary histories. In order to identify the biogeographical connections that exist within a group of organisms, it is necessary to understand the evolutionary relationships of the members in this group. To do this, we will focus on the phylogenetic links within the clades under consideration. Over the past few decades, molecular phylogenetic analyses have become increasingly numerous and complete. Beyond the simple evolutive relationships that they provide, these trees make it possible to construct paleobiogeographical hypotheses. However, these models, which are based on atemporal topologies, require calibrations provided by fossils. As we shall see, the schemas obtained using these calibrations are often inconsistent with paleontological data, but the differences seem to be decreasing in recent publications. Often – too often even – this incompatibility between scenarios based on molecular analyses and those based on fossil records is attributed to the imperfection of the latter. As the saying goes, "The absence of evidence is not evidence of absence", which is true, but only within certain limits. Presuming that the absence of a taxon predicted by a model in a given time and space is exclusively linked to a lack of paleontological data is not necessarily the most prudent hypothesis. For example, tracing the origin of ostariophysans to the Paleozoic, as some calibrated molecular phylogenetics suggest, is in complete contradiction with what we know from fossils. If the fossil record is so incomplete, we would have to abandon the possibility that it could teach us anything about the evolutionary history of these animals. And so this book would be obsolete. Facing a similar problem concerning the evolutionary history of osteoglossids fishes, Kumazawa and Nishida [KUM 00] state: "We interpret this apparent discrepancy [between molecular and fossil evidence] to be indicative of the paucity of osteoglossiform fossil records rather than the inferiority of our molecular time estimates" (p. 1876). Following instead from [FOR 10, p. 237] "we beg to differ" on this point, for reasons that will be explained further on in this work.

Fossils of freshwater fishes are relatively rare. Paleontological sites containing a large diversity of bony fishes, in both the number of individuals and the number of species, are known in the Mesozoic and Cenozoic eras, but they are primarily of marine origin. They are often concentration and conservation Lagerstätten, of which the most well-known are as follows: for the Triassic period, the Monte San Giorgio on the Swiss-Italian border and Luoping in the Sichuan province of China; for the Jurassic period, Solnhofen in Germany and Cerin in France; for the Cretaceous period, Haqil, Hgula and En Namourra in Lebanon; and for the Cenozoic, Monte Bolca in Italy. However, there are some Mesozoic and Cenozoic Lagerstätten that preserve flora and fauna from fresh water environments. They generally correspond to ancient lakes. The fish assemblages that they contain are less diverse than the Lagerstätten of marine origin. Without describing them in detail in this section, they include the following: the Triassic and Jurassic formations of the Newark basin in the United States, the Jehol biota of the Lower Cretaceous in China and the Eocene Green River Formation in the United States. In addition to these exceptionally preserved sites, in which the specimens are often preserved whole and anatomically connected, there are several freshwater paleontological deposits where the fossils are preserved as disjointed pieces. This kind of conservation corresponds to continental environments where energy was higher than it is at the bottom of a lake. This can include rivers, deltas or lagoons. In these environments, the fossils are usually concentrated in the calmest areas, such as the turns of a river or the channel bottoms of a delta. Lastly, the majority of the fossil sites of continental origin are generally more difficult to date than marine sites since, unlike marine sites, continental sites only rarely contain precise fossil markers.

Despite the relative rarity and the low diversity of paleontological sites preserving freshwater ichthyofauna compared to marine sites, the record is sufficient to trace the broad evolutionary steps of groups of fish typical of these environments. On a long time scale and without accounting for recent anthropic action, it should be noted that the biodiversity of freshwater fishes, as a whole, displays a very substantial diversification beginning in the middle of the Mesozoic era. A recent study [GUI 15a] about the modalities of this diversification, based on the fossil record and data using information extracted from morphological and molecular phylogenetics, shows that the diversity of freshwater fishes as a whole has increased exponentially during the past 150 million years. This tendency contrasts with the dynamic of increase in the global diversity of marine fishes, which seems to reach finite capacity over time. This difference in the increase of the biodiversity of fishes depending on their environment is discussed in section 5.2.1.

It is now a question of examining how this extraordinary diversification occurred. We will begin 250 million years ago in the Triassic, a period in which the clades that made up the fauna of freshwater fishes were almost entirely different from the current clades. Next, we will travel to the Jurassic, which probably contains the least amount of information about ichthyofauna, and then on to the Cretaceous, during which the transition from "primitive" to "modern" ichthyofauna took place. Since the beginning of the Cenozoic, the fauna of freshwater fishes has been essentially composed of extant families, even extant genera, and studying them consists of analyzing the establishment of modern fish assemblages. Above all, we are interested in the introduction of freshwater wildlife on a continental scale and we will not examine the factors responsible for the current intracontinental layouts that have resulted from the climatic changes in the Neogene, such as the aridification of North Africa and the Arabian plate, the closure of the Tethys Sea and the Messinian Event, the ice ages and especially the retreat of large glaciers from Europe and North America as well as the implementation of a monsoon climate in Asia. Additionally, only the data based on skeletal fossils are taken into account. I am aware that information gathered from otoliths could provide slightly different scenarios for some lineages (the first occurrences would be pushed back in time in some cases). However, the approach based on the study of otoliths is distinct from this work and based on different literature.

Freshwater Environments and Fishes

1.1. Environments of freshwater ichthyofauna

1.1.1. *Areas and volumes of freshwater environments*

Freshwater environments, which include all continental aquatic environments, are difficult to define (Figure 1.1). As the name indicates, freshwater is characterized by low concentrations of salt as opposed to seawater (the term "sweetwater" is also used as a synonym for "freshwater"). Generally, freshwater is defined as containing less than 0.05% of dissolved salts. Freshwater only accounts for about 2.5–2.75% of all water on Earth. However, 1.75–2% of freshwater is frozen in polar ice caps and glaciers as ice, and 0.5–0.75% exists as groundwater. This leaves about 0.01% of freshwater on the surface where fish can live. Today, nearly three-quarters of freshwater is concentrated in the Great Lakes region of Africa, the Great Lakes in North America and the Baikal Lake in Siberia. Certain continental aquatic environments do not contain freshwater, but still share some characteristics that connect them to the environments we are interested in, despite them having higher salt concentrations. These include sabkhas, which are depressions in hot climates from which water evaporates which increases the salt concentration.

Brackish water has an intermediate salt concentration between freshwater and marine water, between 0.05 and 3%. There are several types of brackish environments today. The most substantial in terms of area are estuaries and river mouths subject to tidal influence. Large rivers discharge such substantial quantities of water into the ocean that brackish or even freshwater ecosystems can extend considerable distances out into the sea. These environments could form freshwater connections across narrow oceans when two rivers face one another. This was the case during the Cretaceous and at the start of the Paleogene when South American and African rivers discharged their freshwater into the proto-Southern Atlantic Ocean. The introduction of freshwater into the marine environment allowed freshwater fishes to spread from one river basin to the next by following the coast [SCH 52]. However, this large outflow could also have acted as a barrier that prevented the spread of marine species along the coast [ROC 03]. In present-day brackish environments, the salinity of mangroves varies according to tidal influence. Modern mangroves have existed since the lower Eocene [PLA 01] but ecosystems with features comparable to mangroves must have existed in the Mesozoic, or at least since the Cretaceous [VUL 08]. Deltas, another kind of brackish environment, form on accumulations of sediment located on the border between marine and freshwater areas. These three types of environments, river mouths, mangroves and deltas, are all characterized by elevated rates of sedimentation as well as abundant and diverse life. These two conditions are favorable for the formation of rich fossil deposits (Figure 1.1). It is very probable that the assemblages of fossilized vertebrates originating from this kind of environment are overrepresented in the fossil record. However, these environments are the most difficult to interpret from a paleoecological point of view because they present a mix of non-native elements from freshwater environments upriver, potentially marine elements brought back to the coasts, and typical elements native to brackish environments. Finally, some lakes and seas can be considered brackish due to high and low salinity, such as the Caspian Sea (which is a lake) and the Baltic Sea, respectively.

One central feature of freshwater environments is the kinetic energy of water in movement (Figure 1.1). If the energy is low and the water is stagnant, we are in the presence of lakes or ponds. These are called lentic environments. If the energy is strong and creates a significant current, we are in the presence of torrents and rivers, which are lotic environments. Lotic environments only represent about 1.2% of open freshwater on the Earth's surface, whereas the remaining 98.8% are lentic environments. This physical feature has a considerable effect on the fish living in this environment as well as on the way the fossils that accumulate there are preserved: although the fossils of fish that lived in ancient lakes are often preserved whole, the fossils of fish that lived in faster moving water are often dislocated and fragmented. An example of this can be observed in the contrast between the fossil records of two groups of otophysan fishes in the Cenozoic in South America: while the characiforms, which lived primarily in rivers, are represented by isolated teeth, the siluriforms, which lived in lentic habitats, are more numerous and represented by more complete fossils. In addition to the difference in habitat, the bones of the siluriforms are generally thicker and denser than those of the characiforms, which increases the probability of fossilization [LUN 98].

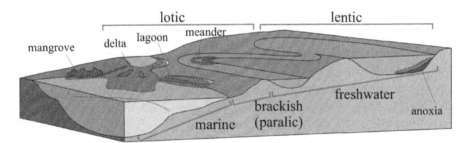

Figure 1.1. *Schematic representation of aquatic environments with an emphasis on freshwater and brackish environments. The zones in red are the most favorable areas for the fossilization of freshwater fishes. For a color version of the figure, see www.iste.co.uk/cavin/fishes.zip*

1.1.2. *Mesozoic and Cenozoic freshwater environments and their fish fauna*

It is difficult to fully map the nature and diversity of freshwater environments in the Mesozoic and Cenozoic because the sedimentary record in these environments is rather incomplete. Due to the absence of polar ice caps during a large part of the Mesozoic and the Paleocene, sea levels were on average much higher than they are today. Epicontinental seas have covered vast areas on all continents several times, spanning across Africa (the Trans-Saharan Seaway), North America (the Western Interior Seaway), Eurasia (the Turgai Strait) and South America. These high sea levels created many brackish ecosystems in paralic environments or interior seas where vast, shallow stretches were subject to extreme variations in salinity. The abundance of these intermediate and mixed environments with an increased potential for fossilization explains why many Mesozoic paleontological sites containing fossil assemblages of fishes are so difficult to define ecologically. For example, in 1952, Schaeffer believed that, with the exception of some formations from the Triassic, no continental formation from the Mesozoic contained any "reasonably representative" fish fauna. Based on our current understanding, we know that this opinion is extreme and certainly exaggerated, as our work will demonstrate.

While the present-day families and genera appeared gradually in the Cenozoic, the composition of the ichthyological assemblages often reveals information about the paleoenvironments. Deducing a type of environment from only a single taxon's autoecology risks falling into a circular reasoning, but this analysis is, however, valuable if it incorporates a set of signals that point toward the same environment. For example, Gaudant [GAU 82] used this method to describe the conditions of formation for three brackish deposits from the Cenozoic: Eocene gypsum of Montmartre, Oligocene gypsiferous marl of Aix-en-Provence and Messinian evaporite deposit in Italy. According to the composition of the fish faunas, the first two assemblages are continental but close to the coast and the third corresponds to an environment with a variable salinity with marine influences.

1.1.3. *Climates and sea levels*

The global climate has varied considerably in the last 250 million years. This is fundamental to our understanding of the evolutionary history of freshwater fishes because even if physical connections between distinct geographical regions allow dispersal events, organisms must still find suitable climate conditions in the region to be colonized. Throughout a large part of the Mesozoic, specifically in the Cretaceous and much of the Paleocene, the average global temperature was higher than it is today (see Figure 1.2). Peak temperatures were reached in the Cenomanian 100 million years ago and during the Paleocene–Eocene Thermal Maximums 56 million years ago. These peaks were followed by long hot periods, in particular a thermal maximum in the lower Eocene.

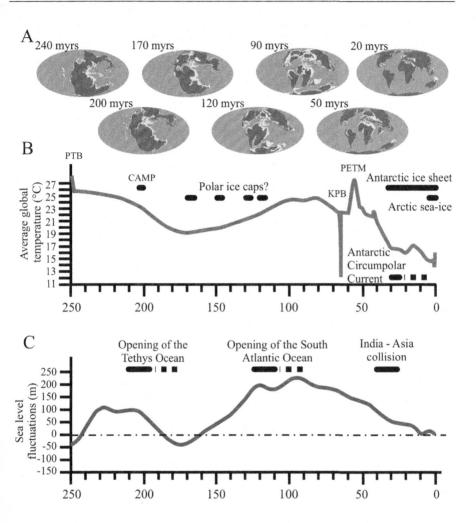

Figure. 1.2. *a) Paleogeographic outline in the Mesozoic and Cenozoic. Ages in millions of years; b) Average temperature in the Mesozoic and Cenozoic (according to [SCO 15]) associated with various global events. CAMP, Central Atlantic magmatic province; KPB, Cretaceous–Paleogene boundary; PETM, Paleocene–Eocene thermal maximum; PTB, Permian-Triassic boundary; c) Sea level associated with a few major tectonic events (according to [VÉR 15]). For a color version of the figure, see www.iste.co.uk/cavin/fishes.zip*

Sea level height, or eustasy, has a considerable influence on the diversity of freshwater fishes (see Figure 1.2). Sea level directly affects the degree of fragmentation of land surface by creating barriers that are impassable for many freshwater fishes. These barriers create islands inside continental masses, but also separate islands from continents and sometimes separate the continents between them. Sea level height certainly plays a role as important as, or even more important than the simple tectonic distribution of continental masses in the distribution of freshwater ichthyological fauna. This factor should be considered when examining paleogeographical maps, which generally do not show sea level. For example, looking at a current world map, it is easy to forget that only a few thousand years ago, although the layout of the tectonic plates was similar to today, many of the south-eastern Asian islands were still attached to the section of the Eurasian block known as Sundaland, Papua New Guinea was still connected to Australia, and a continental connection, Beringia, linked North America to Asia. Another effect of sea level variation pertains directly to the colonization of the continental aquatic environment. Marine incursions promote the transition of marine fishes to a freshwater lifestyle, as has been demonstrated during the incursions in the Miocene in the Neotropical zone [LOV 06]. Some of the freshwater forms in this region that originated in the sea currently include ophichthids, engraulids, pristigasterids, batrachoidids, belonids, hemiramphids, sygnathids, sciaenids, mugilids, gobiids, achirids and tetraodontids [LOV 06]. Current sea levels are low relative to this time period because the polar ice caps retain huge quantities of water, but levels were even lower during the last great ice ages when ice caps were more substantial. During the Mesozoic, on the other hand, sea levels were much higher on average, sometimes by more than 200 m. While a high sea level fragmented the continental environment by creating new inlets, a low sea level increased the surface of the low-lying areas and encouraged freshwater fishes to spread from one hydrographic basin to another. Although it is difficult to precisely determine their number and duration, periods of low sea level may have existed during the Mesozoic, brought about by colder periods and the possible presence of ice caps. Such periods are suspected during the Bajocian-Bathonian, Tithonian, Valanginian and the lower Aptian [PRI 99, PUC 03].

1.1.4. Biological crises

The period under study in this book is marked by several mass extinctions. The first, the greatest mass extinction of life in the entire Phanerozoic, occurred at the transition between the Permian and the Triassic. The second occurred at the Triassic-Jurassic boundary, and the third occurred at the Cretaceous-Paleogene boundary. There are many hypotheses that attempt to explain these three events. We will not discuss them here, but there is a consensus that these events occurred rapidly on a geological scale and were caused by extraordinary and catastrophic events. It is generally accepted that there were sudden, violent and global transformations of the environment related to enormous volcanic eruptions in Siberia at the end of the Permian and in the Central Atlantic magmatic province at the end of the Triassic, and the effects of a meteorite impact that may have coincided with volcanic activity on the Deccan plateau in India at the end of the Cretaceous.

The effects of these events on freshwater fishes are discussed in Chapter 5 (section 5.3.2).

1.2. What is a freshwater fish?

The initial decision to define categories for freshwater fishes was driven by the desire to conduct biogeographical studies. Several groups of fish served as models for patterns applicable to all continental fauna. The first attempts are old and based on a fixed geophysical framework. These models often allude to links between the continents, sometimes without specifying the nature of them (for example [REG 22] about ostariophysans). The wide diversity of the fishes' lifestyles and the ability of many of them to live in environments with variable salinity means that defining clear categories to distinguish types of freshwater fishes is problematic. The classifications generally consider a succession of categories that range from strictly freshwater fishes, or "primary freshwater fishes" [MYE 49], to primarily marine species that occasionally appear in freshwater. Primary freshwater fishes are considered to be one of the most reliable organisms for tracing biogeographical links because, theoretically, even the smallest inlets become impassable obstacles for them. In a noteworthy article about the distribution of freshwater fishes during the Mesozoic, Patterson [PAT 75] further refined the definition of primary freshwater fishes by introducing the terms "archeolimnic," which includes fishes that have inhabited freshwater since the origin of their group, and "telolimnic," which includes fishes whose ancestors had not necessarily always lived in freshwater. As we will see later, it does not seem that a single clade at a supraordinal rank would have been archeolimnic during the Mesozoic and Cenozoic eras. In fact, the most basal representatives of all of the greater freshwater clades were marine. Transitions to freshwater environments generally occurred after the phylogenetic individualization of these clades. In terms of families, Patterson [PAT 75] only identified one archeolimnic family, hiodontids, among the 15 present-day families that he traced back to the Mesozoic. Patterson also believed that a fossilized taxon could be exclusively and necessarily considered primary freshwater if it belonged to a clade whose current representatives were all primary freshwater species. On the other hand, he was suspicious of the practice of reconstructing a taxon's lifestyle based on environmental factors revealed by the sedimentology of the site where it was found. In his view, the presence of a taxon in a continental environment did not necessarily imply its absence in marine environments. The decision to seek a strict definition and limit to the categories of fish lifestyles in order to use them for biogeographical reconstruction is justifiable to avoid an overinterpretation of paleontological data. However, there are several examples of environment transfers in recent history as in the distant past, especially transfers from a marine environment to a freshwater environment, that are revealed by the study of paleontological sites' sedimentological context. It is possible to determine the likelihood of theoretical ancestral environments of clades by examining the environments occupied by different members of the clade. This probabilistic approach renders a strict definition of environmental categories more or less ineffective since, as we will see, each clade has a unique history that is often littered with environmental transfers. This history is not immediately clear from living forms, and the input of the fossil record is necessary. The categories defined by Myers and especially Patterson, however, remain very useful if they are not taken too literally.

Among Myers' intermediate categories between strictly freshwater and strictly marine forms ("secondary", "vicarious", "complementary", "diadromous"), the "vicarious" category represents a particularly interesting group when we examine the present-day distribution of freshwater forms. These freshwater fishes are derived from marine clades that recently invaded continental environments. They are especially present in regions of the world where primary freshwater fishes are absent, such as Australia, New Guinea, large parts of Madagascar and the Philippines, and thus they represent, on a global scale, an example of "opportunistic" occupation of environments free from groups that are dominant elsewhere. During part of the Mesozoic and possibly into the Cenozoic era, Europe was often composed of islands isolated from the large neighboring continental masses, and the region was essentially occupied by these "vicarious" fishes (cf. 5.2.2.2).

1.2.1. *Air breathing*

Fish extract oxygen from water through their gills. However, several species – 374 species belonging to 125 genera, according to Graham [GRA 97] – are capable of breathing air in varying degrees. This ability is generally limited to freshwater fishes living in oxygen-poor habitats, coastal fishes that have developed the habit of temporarily leaving the water or fishes that are subject to emergence events in temporary tidal pools. The following information about current forms is largely borrowed from a summary on the subject published by [GRA 97]. The organs that allow for air respiration in fishes are varied and concern different epithelia of the digestive tract (mouth, pharynx, esophagus, stomach, intestine), the gills or the spaces associated with the branchial cavity (pharyngeal pouches, suprabranchial chambers) or even the skin. The organs that are most commonly used for this function are the lungs and the air bladder. These two organs are distinguished by their position in relation to the digestive tract, which is ventral for the former and dorsal for the latter, as well as by the position of the tube that connects these organs to the digestive tract, and finally by the presence of pulmonary circulation. The relationship between the air bladder present in most bony fishes and the lungs present in some bony fishes as well as tetrapods has been known for a long time. Darwin [DAR 59] considered the transformation of the air bladder into lungs to be an example that illustrated the evolution of one organ within a lineage. Today, however, it appears that the direction of transformation is the reverse of Darwin's proposal. It is the air bladder, whose primary function pertains to the buoyancy of fishes, that is derived from a lung whose function was originally respiratory and/or also related to buoyancy. While the presence of lungs in non-osteichthyan fishes like placoderms is still debated [DEN 41, JAN 07, GOU 11], this organ was certainly present in the common ancestor of the osteichthyans.

Direct evidence of the existence of lungs is extremely rare in fossils, but the distribution of this organ in current forms informs us of its presence and its possible structure in extinct species. In the case of *Latimeria*, the modern coelacanth, a vestigial lung has recently been observed [CUP 15]. It is proportionally well developed and potentially functional at the embryonic stage. This vestigial lung has several small calcified plates on its surface that we find, in a very developed form, in various fossil species of coelacanths and in particular in Cretaceous mawsoniids. The positioning of these plates in *Axelrodichthys* leads us to believe that they played a role in the volumetric variations of the lung while breathing air, facilitating ventilation like a bellows [BRI 10].

The air breathing of lungfish has been the subject of many studies, in particular in the genera capable of estivation *Protopterus* and *Lepidosiren*. The Australian genus *Neoceratodus* does not have mandatory air breathing and does not estivate, contrary to the two other modern-day genera. The ability to estivate has been attributed to the Permian genus *Gnathorhiza* following the discovery of several burrows containing the remains of this lungfish [ROM 54]. The lungfish burrows, probably used for estivation, are rarer in the Mesozoic, but they are not absent. We know they exist in the Triassic [GOB 06] and the Cretaceous [ORS 07] of the United States, the lower Cretaceous of Denmark [SUR 08] and the upper Cretaceous of Madagascar [MAR 12]. The identification of the ability for extinct forms to breathe air and estivate relies on, aside from rare direct evidence, phylogenetic links between extinct forms and modern forms. If we consider the relatively recent separation of *Neoceratodus* from the *Protopterus* + *Lepidosiren* set, as is generally accepted, then the obligatory air respiration of *Protopterus* + *Lepidosiren* is a characteristic that appeared recently, in the Cretaceous or Jurassic at the earliest. However, a recent phylogeny [KEM 17] revived the old idea of a relationship between *Protopterus* + *Lepidosiren* and *Gnathorhiza*, a lungfish that was relatively common in the Permian. In this schema, the majority of Mesozoic taxa are more closely related to the *Protopterus* + *Lepidosiren* lineage than to the Neoceratodus lineage. As these extinct forms were discovered in freshwater deposits, it is likely that they already had the ability to breathe air.

Among actinopterygians, the ability to breathe air through a lung or an air bladder exists today in the most primitive forms. The most basal order of actinopterygians today, Cladistia or Polypteriformes, have lungs with a structure comparable to those of tetrapods. In the living Holostei, gars and bowfin, the organ corresponds to an air bladder (located dorsally) with a respiratory function. This is also the case in many osteoglossomorphs and the elopomorph *Megalops* (a second elopomorph, the eel *Anguilla*, is capable of breathing air, but this respiration occurs through the skin and possibly the pneumatic duct associated with the digestive tract, not using a lung or air bladder). The other living actinopterygians capable of breathing air through an air bladder include the gonorynchiform *Phractolaemus*, some genera of characiforms and siluriforms, one electric eel and the esocoid *Umbra*. As highlighted by Graham [GRA 97], it is likely that this is a primitive characteristic for actinopterygians. It would be inherited from their common ancestor with the sarcopterygians and preserved in all basal groups until the elopomorphs (Polypteriformes, Holostei, Osteoglossomorpha, Elopomorpha). It is even possible that this characteristic would have been preserved in some otomorphs (that is the clupeomorphs and the ostariophysans) but this is unlikely because the function is absent from clupeomorphs as a whole and remains rare in ostariophysans despite the extreme diversity of this group in freshwater. It is likely that this capacity reappeared incidentally in some ostariophysans. According to this schema, which is to say the existence of lungs or an air bladder with a respiratory function in non-clupeocephalan actinopterygians, this organ and function should be present in several extinct groups, especially in freshwater. We can infer that in Mesozoic ginglymodes, especially *"Semionotus," "Lepidotes"* and related freshwater forms, Cretaceous amiids and sinamiids, basal freshwater teleosts such as archeomenids, siyuichthyids, some aspidorhynchiforms, ichthyodectiforms, and freshwater elopomorphs were all capable of air respiration to a greater or less extent.

Several other actinopterygians can breathe air using different organs with epithelia or directly through the skin. These specializations can be found in the astonishing small salamander fish of Australia (*Lepidogalaxias salamandroides*), the galaxaiids of the Southern Hemisphere, gobiosocids, cyprinodontiforms, perciforms like several blennies and gobies (including the infamous mudskippers that slide along the mud beaches in mangrove forests), channiforms and rare synbranchiforms. The majority of these adaptations are present in intertidal or coastal species that must be able to cope with exposure to air when tides are low or when they voluntarily leave the water. Given the diversity of the organs involved in this function and the phylogenetic distance that separates the taxa in question, these specializations certainly correspond to distinct adaptations. To my knowledge, the fossil record of the Mesozoic and Cenozoic do not inform us about the origin and evolution of these structures.

1.2.2. *Weberian apparatus*

The Weberian apparatus is a remarkable feature shared by all otophysans (Figure 1.3). It is a sensory organ constructed out of modified vertebrae and their associated components that is located just behind the skull. The organ transmits sound vibration from the air bladder to the inner ear, somewhat similar to the tiny bones in the middle ear of mammals. It can be assumed that this organ is one of the keys to the evolutionary success of otophysans in freshwater because, in the many freshwater environments where the water is murky, audition is a significant advantage over vision. The structure of the Weberian apparatus in different otophysan clades and the phylogenetic marker that it represents were discussed by Fink and Fink [FIN 81, FIN 96], Grande and de Pinna [GRA 04] and Chardon *et al.* [CHA 03] in particular. The evolutionary origin of this structure as illustrated by paleontological data is discussed in Chapter 4 (section 4.15.2.2).

Dorsal view

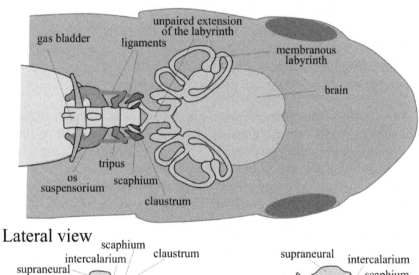

Lateral view

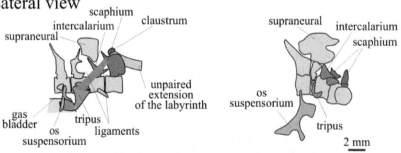

Figure 1.3. *Schematic representation of the Weberian apparatus in otophysans from the dorsal view and the lateral view (according to various sources). Bottom right, the "primitive" Weberian apparatus of* Chanoides macropoma *from the Eocene of Monte Bolca in Italy (according to [PAT 84]). For a color version of the figure, see www.iste. co.uk/cavin/fishes.zip*

1.2.3. *Pharyngeal dentition*

The diets of freshwater fishes are incredibly varied. In the most diversified group, the otophysans, we observe the full range of sources of nutrition and diets. It stretches from the algivorous sucking loach of the family gyriocheilids that scrapes algae off hard substrate to the carnivorous piranha. Among the otophysans, the cypriniform fishes are characterized by the absence of teeth in the mouth but the presence of a very developed pharyngeal dentition. This structure is a functional jaw located at the back of the orobranchial cavity, behind the primary jaw formed by the mandible and the maxilla/premaxilla. The pharyngeal jaw has an important role in processing food swallowed by fish that only use their primary jaw to draw in food. The teeth on the pharyngeal jaw have characteristic forms which, when they are found fossilized in sediment, make it possible to identify them taxonomically. The pharyngeal jaws of cichlids are comparable to those of cypriniforms. In cichlids, the inferior pharyngeal jaw, made up of fused ceratobranchials, has specific bony morphology and dental features according to the taxa. These components also have a certain phenotypic plasticity, which means that the form of the bony structure bearing the teeth and the teeth themselves change shape as the individual grows, depending on the diet adopted (cf. for example, [MEY 87, BOU, 02, MUS 11]). This plasticity could be one of the factors at the origin of species flocks that have evolved in this family in the East African Great Lakes (cf. 5.2.2.2).

Another large clade of freshwater fish, osteoglossomorphs, is characterized by a functional jaw between teeth located on the palate and the tongue. In this group, a very specialized form from the upper Cretaceous in North Africa, *Palaeonotopterus*, has palate teeth that are completely fused to the inside of a very dense oval plate that faces its lower counterpart to form a sort of nutcracker [MEU 13].

1.2.4. *Electric fishes*

Electroreception is the ability to perceive low-frequency electric fields using electroreceptive organs derived from the lateral line, an organ that was originally a mechanoreceptor. Electroreception is present in many primitive fishes, such as chondrichthyans, coelacanths and sturgeons. This ability seems to have been lost at the level of the common ancestor of all neopterygians, but it coincidentally reappeared in two groups of freshwater fish: Siluriphysi, which includes siluriforms and gymnotiforms, and Notopteroidei, which includes mormyroids and notopterids. The development of this sense in these fishes is considered an adaptive response to the lack of visibility in the turbid freshwater environments in the same way that the Weberian apparatus in otophysans allows for better auditory perception to compensate for the lack of visibility. In addition to low-frequency electroreception, the development of organs that generate weak electric shocks and electroreceptors that are capable of perceiving the corresponding high frequencies have been observed in two lineages within these two clades, the gymnotiforms and the mormyroids. In these freshwater fishes from South America and Africa, respectively, the new sensory system allows for electrocommunication and an effective electrolocation based on perceiving distortions of the electric field resulting from the presence of an obstacle in the immediate environment. These fish with weak shocks are distinguished from fish with strong shocks, like the electric eel, *Electrophorus electricus* (a South American gymnotiform) and *Malapterus* (an African siluriform), which produce shocks to stun their prey or defend themselves. *Malapterus* can produce shocks of 350 V and *Electrophorus* close to 700 V. Gymnotiforms and mormyroids display surprising convergences in their general morphology and behavior.

The fossil record of gymnotiforms and mormyroids is sparse but the evolutionary history of these two lineages was reconstructed in part by Lavoué *et al.* [LAV 12] based on a molecular phylogenetic analysis. The authors found a remarkable similarity in the history of the two clades, especially in the chronology of their evolutionary history. The origin of electroreception appeared in the two lineages in the Early Cretaceous, about 20 million years before the ability to produce an electric field, at the start of the Late Cretaceous. Lavoué *et al.* noted that these events occurred during the last phases of separation between Africa and South America, the continents where these fishes now live. The authors suggest that specific, probably climatic, environmental conditions may be at the source of the selection pressures responsible for these very particular parallel adaptations.

2

Assemblages of Freshwater Fishes in the Mesozoic

2.1. Triassic

The Triassic, which occurred from 252 to 201 million years ago, was a period when the continental grouping was at its most intact. Almost all of the current continents were grouped within Pangaea and surrounded by Panthalassa, an immense ocean. Pangaea was largely divided by an ocean open to the east that separated Laurasia in the north from Gondwana in the south. This ocean was composed of two distinct ocean floors, the Paleo-Tethys in the north and the Neo-Tethys in the south, which were separated by the Cimmerian continents. The Triassic saw the closure of the Paleo-Tethys by subduction under Laurasia as well as the opening of the Neo-Tethys. At the end of the Triassic, the Central Atlantic magmatic province marked the beginning of the opening of the central Atlantic. The global climate was hot and very continental due to the particular layout of the land masses. The approximate geographic location of the main sites containing freshwater fishes is presented in Figure 2.1.

2.1.1. *Australia*

Three assemblages located in the Sydney basin dating from the Lower Triassic (Gosford) and the Middle Triassic (Brookvale and St. Peters) yielded relatively varied fauna primarily composed of primitive fishes belonging to groups that were already present in the Paleozoic. They were initially studied by Arthur Smith Woodward and Robert Thompson Wade.

The assemblages mostly contain "paleopterygians" in the sense used by Romano *et al.* [ROM 16] (*Palaeoniscum, Myriolepis, Apateolepis, Agecephalichthys, Megapteriscus, Belichthys, Mesembroniscus, Macroaethes*), some redfieldiiforms ("*Dictyopyge*", *Brookvalia, Phlyctaenichthys, Geitonichthys, Molybdichthys, Scizurichthys*) and some "perleidiforms" (*Cleithrolepis, Procheirichthys, Manlietta, Pholidopleurs*), many of which are endemic. The saurichthyiforms (*Saurichthys*) and the neopterygians (*Peltopleurus,* "*Semionotus*", *Promecosomina, Enigmatichthys, Belonorhynchus,* "*Pholidophorus*") are also present but not very diversified [LÓP 04, ROM 16]. Two of these locations also contain the lungfish *Gosfordia*. The Knocklofty Formation from the Lower Triassic in southern Tasmania yielded some fish remains as well as the dental plates of the lungfish "*Ceratodus*" *gypsatus*, some fragments of a possible coelacanth, the "paleoniscoid" *Acrolepis*, the "perleidiform" *Cleithrolepis* and the ubiquitous genus *Saurichthys* [DZI 80].

Late Triassic (230 myrs)

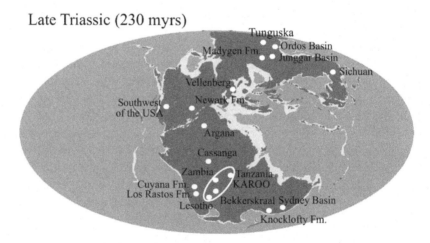

Figure 2.1. *Main sites and formations from the Triassic that yielded faunas of freshwater fishes positioned on a paleogeographic map of the Late Triassic. For a color version of the figure, see www.iste.co.uk/cavin/fishes.zip*

2.1.2. Africa

2.1.2.1. The Karoo supergroup

This supergroup is an accumulation of sedimentary rocks that began in the Permian and continued until the Lower Cretaceous. Within the Karoo deposit, the levels richest in fish fossils form the Beaufort Group, which extends from the Upper Permian to the Lower Triassic. The upper part of the group corresponds to fluvial and lacustrine environments. The most diversified ichthyological assemblage dates from the Anisian and comes from Bekkerskraal in the Orange Free State. Like the Australian fauna, it contains lungfish, "paleopterygians" (*Elonichthys, Dicellopygae*), redfieldiiforms (*Atopocephala, Daedalichthys, Helichthys*) and "perleidiforms" (*Meidiichthys, Cleithrolepidina, Hydropessum*), but no neopterygians [MUR 00, LÓP 04]. The Karoo is also present in Tanzania where *Australosomus, Acrolepis* and other taxa with unclear affinities [MUR 00] were discovered, as well as in Zambia and Lesotho where *Endemichthys likhoeli* [FOR 73] was found. The upper part of the Karoo contains fewer fossilized fishes. The only recorded mention is "*Semionotus*", cf. *Semionotus capensis* from the Stormberg Group in South Africa dating from the Upper Triassic [JUB 75, MUR 00].

In Angola, the Cassanga Series from the Lower Triassic yielded an assemblage of freshwater fishes that were probably preserved in a lake whose level fluctuated in a climate that varied between dry and wet seasons. The fauna contained the paleopterygian *Schaefferius moutai* (initially *Marquesia* but the name was preoccupied), the perleidiforms "*Perleidus*" *lutoensis* and "*Perleidus*" *lehmani* and the neopterygian *Angolaichthys lerichei*, as well as the lungfish *Microceratodus angolensis* [ANT 90]. *Schaefferius* is attributed to the canobiids, which is otherwise restricted to the Carboniferous period. *Angolaichthys* is considered to be included in the semionotids by Murray [MUR 00], but in the figure in [ANT 90], we do not observe any ginglymodian characteristics.

2.1.2.2. Morocco

In the Argana corridor, freshwater fishes from the Upper Triassic were described, including lungfishes [MAR 81, KEM 98], the perleidiform *Dipteronotus* [MAR 80a] and the redfieldiiform *Mauritanichthys* [MAR 80b], as well as a fragment from a coelacanth's mandible.

2.1.3. South America

The ichthyofauna in the Cuyana basin, in Argentina, was recently revised by López-Arbarello *et al.* [ARB 10]. It contains the chondrostean *Neochallaia*, the acrolepid *Challaia*, the genus *Guaymayenia* with unclear affinities and five perleidiforms from the pseudobeaconiids family. Another small lacustrine ichthyofauna was described from the Ladinian in the Los Rastos Formation in the La Rioja province in Argentina [LÓP 06]. The fauna included five species, four of which are endemic to this site (*Gualolepis carinaesquamosa*, *Rastrolepis riojaensis*, *Rastrolepis latipinnata* and *Challaia elongata*). Their affinities within the actinopterygians are unclear but they correspond to basal forms attributed here to the "paleopterygians". *Challaia elongata*, is a member of the acrolepids, a primarily Paleozoic family that has few representatives in the Triassic, with the Argentinian occurrence being the youngest for the Gondwana.

2.1.4. Europe

The Middle Triassic in Germany (Muschelkalk and the lower Keuper) yielded freshwater ichthyofauna including the fossil assemblages from Vellberg in southern Germany. Schoch and Seegis [SCH 16] recently re-examined these faunas. One of the assemblages corresponds to an olighaline lake containing the perleidiforms *Dipteronotus* and *Serrolepis*, "*Gyrolepis*", some redfieldiids, scanilepids, ginglymodians, chondrosteans and polzbergiids, as well as some lungfish and coelacanths. The organisms were preserved on the anoxic bed of a lake with significant vertical zonation.

2.1.5. Asia

2.1.5.1. China

The continental fishes of the Triassic in China primarily come from the intramontane basins of Junggar in the west, Ordos in the center-north and Sichuan in the south [JIN 06]. The Lower Triassic sites contain the redfieldiiform *Sinkiangichthys* and the scanilepiform *Fukangichthys*. The Middle Triassic sites contain the ginglymodian *Sinosemionotus*. Finally, the Upper Triassic sites contain various "paleopterygians", *Saurichthys* and the pholidophorid *Jialingichthys*.

2.1.5.2. Central Asia

The ichthyofauna of Tunguska, located near the river of the same name, developed in a lacustrine environment that probably dates from the Lower

Triassic or potentially the terminal Permian. The fauna was reviewed in 1999 by Sytchevskaya. It primarily contains relict taxa from the Paleozoic era, such as scanilepiforms, perleidiforms, pholidopleuriforms and the genus *Tungusichthys*, classified in its own family: the tungusichtyids. Sytchevskaya [SYT 99] thought that this family could be integrated into the "semionotiforms". However, the published photographs and drawings of this taxon indicate that it is not a ginglymodian and probably not a neopterygian either.

The Madygen Formation in Kyrgyzstan, dating from the Middle-Upper Triassic, (Ladinian Carnian), is a fluvial-lacustrine deposit that yielded the lungfish *Asiatoceratodus sharovi*, some scanilepiforms, "paleopterygians", perleidiforms [SYT 99] and *Saurichthys* [KOG 09]. This fauna is similar to the Siberian fauna in Tunguska, which seems to indicate a constant endemism during a large part of the Triassic. However, the endemic character of the ichthyofauna may be overestimated due to the lack of information from deposits of a similar age in Asia. When we consider higher taxonomic ranks than genera, for that matter, this ichthyofauna fits well in a broader group characteristic of central Asia and northern China, according to [CHA 04].

2.1.6. *North America*

2.1.6.1. *Newark Formation (Upper Triassic–Lower Jurassic)*

At the end of the Triassic, before the ocean that would separate North America from North Africa opened, a series of basins formed and held great lakes, whose configuration is somewhat comparable to the East African Great Lakes today. Like those lakes, the fauna of freshwater fishes presented rapid radiations called "species flocks" (cf. 5.2.2.2) [MCC 87, MCC 96]. A succession of five assemblages of fishes, called zones, was recognized as stretching from the Late Triassic to the Early Jurassic [OLS 82]. The oldest assemblage is dominated by the redfieldiid *Dictyopyge* accompanied by another genus from the same family, *Cionichthys*, and other basal forms such as the perleidiform *Tanaocrossus*. A second assemblage, possibly from the same time as the previous one but present in other basins, is characterized by the coelacanth *Diplurus newarki* as well as the "paleonisciform" *Turseodus* and the ginglymodian *Semionotus brauni*. The following zones include semionotids and date from the Jurassic. These fishes show a wide morphological diversity, which is the source of extensive taxonomic confusion. They are assembled in "groups" that follow after one another and constitute species flocks: the "*Semionotus brauni* group" present in the *Diplurus* zone, the "*Semionotus tenuiceps* group" and the "*Semionotus micropterus* group". In addition to these *Semionotus*, there are some *Redfieldius*, *Ptycholepis* and *Diplurus*.

2.1.6.2. Southwestern United States

In the southwestern United States, a series of formations straddling the Triassic–Jurassic boundary yielded rich freshwater ichthyofaunas. They include the Chinle Formation and the Dockum Group from the Upper Triassic as well as the Moenave and Kayenta Formations from the Lower Jurassic. Milner *et al.* [MIL 06] defined a succession of fish assemblages based partly on isolated remains and partly on complete specimens. The taxa represented are mainly "paleopterygians", redfieldiiforms and "perleidiforms" to which was just added, especially in the Lake Dixie assemblage, some ginglymodians initially attributed to the genus *Semionotus* and recently included in the genus *Lophionotus* [GIB 13a, GIB 13b]. *Hemicalypterus weiri*, from the Upper Triassic in the Chinle Formation, was recently redescribed by Gibson [GIB 16]. She places this form within the dapediiforms, which appear in this study as a sister group of the ginglymodi within the holosteans. This species lived in a lacustrine-fluvial-deltaic environment in a monsoon climate, and according to Gibson, it represents the most ancient example of a typically plantivore actinopterygian. *Chinlea sorenseni* is a mawsoniid coelacanth discovered in the Chinle Formation.

2.2. Jurassic

The Jurassic stretched from 201 to 145 million years ago. The start of the period saw Laurasia move away from Gondwana. In the Late Jurassic, the North Atlantic began to open, separating North America from Europe, and Gondwana broke apart. Western Gondwana, which included South America and Africa, detached from eastern Gondwana, which included Antarctica, Australia and India. The fossil record of freshwater fishes from the Jurassic is the poorest of the entire timeframe under consideration. The approximate geographic location of the main sites containing freshwater fishes in the Jurassic is presented in Figure 2.2.

Middle Jurassic (170 myrs)

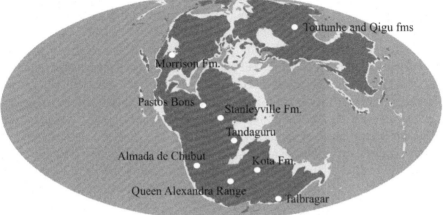

Figure 2.2. *Main sites and formations from the Jurassic that yielded faunas of freshwater fishes positioned on a paleogeographic map of the Middle Jurassic. For a color version of the figure, see www.iste.co.uk/cavin/fishes.zip*

2.2.1. *South America*

The lacustrine assemblage from the Late Jurassic at Almada, in the Chubut province of Patagonia, contains the basal teleost, *Luisiella*, the coccolepidid *Condorichthys* and some basal actinopterygians with unclear affinities. It is considered to be very similar to the fauna in Talbragar in Australia [LÓP 13].

The Pastos Bons Formation in the Parnaiba basin in Brazil was probably deposited in the Late Jurassic in an environment controlled by a fluvial-lacustrine system. It yielded the pleuropholid *Gondwanapleuropholis longimaxillaris* [BRI 02].

2.2.2. *Australia*

The Talbragar site in New South Wales contains a rich freshwater ichthyofauna. The site was formed in a lacustrine environment where mass mortality events occurred following nearby volcanic eruptions [BEA 06]. It was initially described by Woodward at the end of the 19th Century

[TUR 16]. Absolute dating of the site based on an ash level beneath the fossiliferous level gave an age of 151 million years, which is to say Kimmeridgian. The most commonly found fishes are the basal teleolost *Cavenderichthys talbragarensis* (initially included in the genus *Leptolepis*), followed by a coccolepidid and some archaeomenids (*Archaeomaene* and *Madariscus*). Added to this list are *Aetheolepis* and *Aphnelepis*, which were initially given their own family but are now grouped with the archaeomenids [TAV 11], and *Uarbryichthys* [BEA 06]. *Aetheolepis mirabilis* played an important role in Woodward's understanding of the classification of fishes. On the anterior part of its body, this fish had ganoid scales, while on the posterior part of its body, it had cycloid scales. The presence of two types of scales on the same individual that supposedly characterize two very different types of fish challenged the general classification of actinopterygians elaborated by Louis Agassiz in the first half of the 19th Century, which was still largely accepted [FOR 16].

2.2.3. Antarctica

Deposits of lacustrine origin from the Lower Jurassic intercalated within the Kirkpatrick basalts in the Queen Alexandra Range in Antarctica yielded several specimens of the archaeomenid *Oreochima ellioti*. Schaeffer [SCH 72] regarded this species as closely related to Australian archaeomenids, in particular *Aetheolepis* and *Aphnelepis*.

2.2.4. India

According to López-Arbarello *et al.* [LÓP 08], the Kota Formation dates from the Early Jurassic. The paleoenvironment is disputed and some researchers believe it to be freshwater. This formation contains the ginglymodian "*Lepidotes*" *deccanensis* [JAI 83], the neopterygians *Paradapedium* and *Tetragonolepis* that have unclear affinities, a basal teleost attributed to the catch-all genus *Pholidophorus* and the coelacanth *Indocoelacanthus robustus*.

2.2.5. Africa

The Stanleyville Formation in the Democratic Republic of Congo, now dated to the Aalenian-Bathonian (Middle Jurassic), contains a rich assemblage that was initially described by Pierre de Saint-Seine and Edgard Casier and later revised by Louis Taverne. It corresponds to a lacustrine fauna, possibly in a saline lake [TAV 11a], which contains the catervariolids *Catervariolus hornemani*, *Songanella callida* and *Kisanganichthys casieri*, the ligulellid *Ligulella sluysi*, the ankylophorids *Steurbautichthys aequatorialis* and *Songaichththys luctacki*, the ophiopsid *Congophiopsis lepersonnei*, the pleuropholids *Pleuropholis lannoyi*, *Pleuropholis jamotti*, *Parapleuropholis olbrechtsi* and *P. koreni*, *Austropleuropholis lombardi*, the lombardinid *Lombardina decorata*, the signeuxellid *Signeuxella preumonti*, the majokiid *Majokia brasseuri* and the basal teleost *Paraclupavus caheni* [MUR 00, TAV 01, TAV 11a, TAV 11b, TAV 11c, TAV 13, TAV 14a, TAV 14b, TAV 14c]. Finally, the coelacanth *Lualabea*, a probable mawsoniid, is also present.

The Upper Jurassic site in Tendaguru in Tanzania is primarily known for its dinosaurs, but it also contains the remains of ginglymodian fishes. Arratia and Schultze [ARR 99] named this species "*Lepidotes*" *tendaguruensis*. Then, in a general revision of ginglymodian fishes, López-Arbarello [LÓP 12] attributed this taxon to the genus *Callipurbeckia*. The environment when the site formed was comparable to that of the Morrison Formation, which was hot and arid with a rainy season that led to the formation of ephemeral bodies of water.

2.2.6. North America

The Morrison Formation, which covers a large area in the western United States, is well known for its dinosaur fauna. The sediments from the Late Jurassic were deposited in vast alluvial plains with ephemeral bodies of water due to substantial evaporation in a semi-arid climate. The ichthyofauna includes the coccolepidid, *Morrolepis schaefferi*, the "pholidophoriform" *Hulettia hawesi*, an indeterminate amiiform known from isolated vertebrae and a pycnodontiform represented by a tooth fragment, as well as indeterminate teleosts [KIR 98]. Lungfish are represented by four species of *Ceratodus*, with one which was later attributed to the genus *Potamoceratodus* [PAR 10].

2.2.7. Asia

The Toutunhe and Qigu formations are made up of continental sediments from the Middle and Late Jurassic in the Junggar basin in western China.

Ganoid scales indicate the presence of five distinct species attributed to relic groups close to *Ptycholepis* [MAI 03].

2.3. Cretaceous

The Cretaceous period, which occurred from 145 to 66 million years ago, was a long period during which Pangaea continued to break apart. The continents that formed Gondwana and the continents that formed Laurasia continued to separate and reached a maximum distance around the beginning of the Late Cretaceous. From there, the movement would reverse following the opening of the South Atlantic that had begun a few tens of millions of years before, in the Early Cretaceous. North America continued to move away from Europe and intermittently came into contact with East Asia. India detached from Gondwana at the start of the period and moved rapidly northward throughout the Cretaceous. Australia also detached from Antarctica during the Upper Cretaceous. The climate was generally hot and maximum temperatures were reached at the start of the Upper Cretaceous. Sea levels were also extremely high.

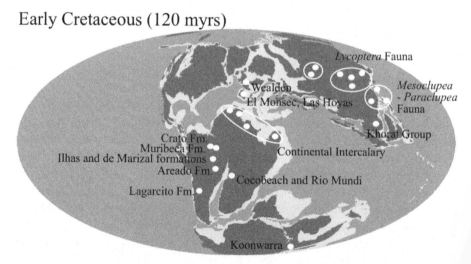

Figure 2.3. *Main sites and formations from the Early Cretaceous that yielded faunas of freshwater fishes positioned on a paleogeographic map of the Aptian. For a color version of the figure, see www.iste.co.uk/cavin/fishes.zip*

The Lower Cretaceous is the period of the Mesozoic that has the richest fossil record of freshwater fishes in the last 250 million years, with the exception of the current period. These assemblages come from the Wealden facies in Europe, Jehol biota and equivalent levels in East Asia, North African Continental Intercalary and equivalent deposits in South America, and from Australia. The Upper Cretaceous contains fewer sites with freshwater fishes, but relatively diversified assemblages are, nonetheless, present in North America, South America, Europe and Asia. The majority of them correspond to lotic environments. The approximate geographic location of the main sites containing freshwater fishes is presented in Figure 2.3 for the Early Cretaceous and Figure 2.4 for the Late Cretaceous.

Late Cretaceous (66 myrs)

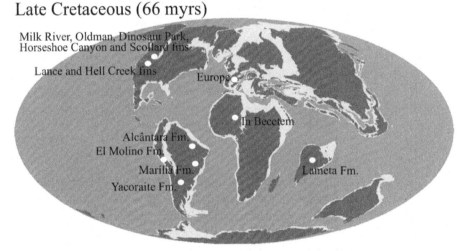

Figure 2.4. *Main sites and formations from the Late Cretaceous that yielded faunas of freshwater fishes positioned on a paleogeographic map of the Maastrichtian. For a color version of the figure, see www.iste.co.uk/cavin/fishes.zip*

2.3.1. *Europe*

2.3.1.1. *Wealden*

The European Wealden is a facies characterized originally in southern England and also found in the north of France, Belgium and Germany. It corresponds to continental environments of flood plains and deltas as well as paralic environments that range from the Valanginian to the Barremian. The Wealden deposits rest on the Purbeckian facies, which corresponds to an intermediary environment between the marine Portlandian level below and the more continental Wealden above. The fishes in this facies were studied by various authors including Woodward on the basis of British fossils [WOO 16]. The updated faunistic list includes the pyncodontid "*Coelodus*" *mantelli*, the possible ophiopsid *Neorhombolepis valdensis*, the caturid *Caturus*, the ginglymodian *Scheenstia mantelli*, the aspidorhynchid *Belonostomus* and the possible elopomorph *Arratiaelops* (= "*Oligopleurus*") *vectensis*. On the European continent, the Wealden facies yielded the iguanodons in Bernissart, Belgium in the 19th Century. These dinosaurs were accompanied by an assemblage of fishes that were described by Traquair [TRA 10]. The ichthyofauna, currently being reviewed, is diversified. It contains, according to a partially updated list, the coccolepidid *Coccolepis macroptera*, the pycnodontid *Turbomesodon bernissartensis*, one or several ginglymodians ("*Lepidotes*"), the oligopleurid *Callopterus insignis*, the pleuropholid *Pleuropholis*, the macrosemiid *Notagogus parvus*, the amiid *Amiopsis doloi*, the "pholidophoriform" *Pholidophorus obesus*, the elopomorph *Arratiaelops vectensis*, the euteleost *Pattersonella formosa*, the possible gonorynchiform *Aethalionopsis robustus* and the osmeriform *Nybelynoides brevis*. As we will see later on, the majority of these species belong to clades that are not primary freshwater fishes.

2.3.1.2. El Montsec and Las Hoyas

El Montsec is a conservation Lagerstätte from the Upper Berrasian–Lower Valanginian located in northeastern Spain. This site had a rich flora and fauna. Although the environment was definitely lacustrine at the time of the deposit's formation, the assemblage of fishes is composed of taxa that are not strictly freshwater. It includes a ginglymodian, the macrosemiids *Notagogus* and *Propterus*, the amiids *Amiopsis* and *Vidalamia*, the caturid *Caturus*, the ophiopsid *Ophiopsis*, the pycnodontid *Ocleodus*, the pleuropholid *Pleuropholis*, the elopid *Ichthyemidion* and the chanid *Rubiesichthys*. In addition to fishes known by their skeletons, a species of the ichnogenus *Undichna* was reported. It corresponds to traces left in the loose sediment at the bottom of the lake by the movement of fishes, probably *Ichthyemidion* [GIB 99].

The Las Hoyas site near Cuenca in Spain is another conservation Lagerstätte that contains a rich fauna of vertebrates preserved at the bottom of a lake [BUS 10]. The area, dating from the Upper Barremian, is slightly younger than the one at El Montsech. According to Buscalioni and Fregenal-

Martínez [BUS 10], the levels of the lake fluctuated due to a subtropical climate characterized by a strong seasonal pattern that caused regular hydric stresses. These cycles were registered in the sediment in the form of limestone laminae. The ichthyological assemblage is similar to that of El Montsec at the family and sometimes genus level. There are ginglymodians attributed to two species of "*Lepidotes*" (more likely a neighboring genus), some macrosemiids (*Notagogus*, *Propterus*), amiids (*Amiopsis*, *Vidalamia*), chanids (*Gordichthys*, *Rubiesichthys*), a pleuropholid (*Pleuropholis*) and some pycnodontids (*Stenamara*, *Turbomesodon*) [POY 95, POY 02, POY 04]. There are also some specimens of the coelacanth "*Holophagus*" [POY 98]. Like at El Montsec, the ichnogenus *Undichna* is represented at Las Hoyas, but by another ichnospecies that corresponds to traces left by pycnodonts [GIB 99]. Strontium isotope analyses that were carried out on the sediment and fossils show that it was clearly a freshwater environment [POY 08]. It is the first proof that pycnodontids and coelacanths occupied freshwater environments in the Cretaceous.

The presence of fishes that do not belong to primary freshwater taxa in Las Hoyas, El Montsech and in the British and Belgian Wealden probably indicates a certain isolation of European islands in the Lower Cretaceous. Like existing island situations, these freshwater environments were not favorable to the evolutionary radiation of primary freshwater taxa, but did "capture" marine forms (cf. 5.2.2.2).

2.3.1.3. The terminal Cretaceous

During the Late Cretaceous, Europe was composed of a series of islands with various biogeographical links. The easternmost islands presented faunistic links with Asia while the more southern islands showed connections with North America and western Gondwana. The diversified fauna of dinosaurs and other tetrapods has been identified in the deposits located on several of the ancient islands, but the assemblages of freshwater fishes are still not very well known and not very diversified. This low diversity could be explained by a lack of sites conducive to preservation, but it may also have been low in reality, as can be observed on certain islands today. The majority of European deposits correspond to fluvial environments where the energy was relatively strong. Consequently, the skeletal elements are preserved in an isolated and fragmented way. The most common remains are ganoid scales associated with teeth and vertebrae attributed to lepisosteid fishes. They have been found in Spain, Southern France, Hungary and

Romania [CAV 96, GRI 99, CAV 99b, SZA 16]. We also know of isolated dentitions attributed to phyllodontids in Spain and France [CAV 99, LAU 99] and, more unexpectedly, remains of a mawsoniid coelacanth that was recently named *Axelrodichthys maegadromos* [CAV 05, CAV 16]. Recently, Blanco *et al.* [BLA 17] described assemblages from the Maastrichtian in different areas of northeastern Spain that correspond to coastal environments and alluvial plains. They found elements that are normal for these levels, such as gars and phyllodontids and, for the first time in the Upper Cretaceous in Europe, an osteoglossiform fragment. They also noted isolated fragments attributed to *Belonostomus* and some cypriniforms. The presence of the latter taxa, represented by isolated teeth and a fragment of a branchial arch attributed to a cypriniform, must be confirmed with more distinctive skeletal elements.

2.3.2. Australia

2.3.2.1. Koonwarra

This site from the Aptian in the State of Victoria, discovered in 1961, corresponds to a lake, at the bottom of which a rich flora and fauna of fish and insects was preserved. Many of the insects belong to groups characteristic of cold environments, which fits with the high paleolatitudinal position of the site. The ichthyofauna includes the coccolepidid *Coccolepis woodwardi*, the archaeomenid *Wadeichthys oxyops*, the basal teleost *Waldmanichthys koonwarri* and the teleost with unclear affinities, *Koonwarria manifrons* [WAL 71, SFE 15].

2.3.3. Asia

2.3.3.1. The East Asian Great Lakes

Deposits from the Early Cretaceous in Liaoning and neighboring regions make up the Jehol biota. They are mainly located in the Yixian Formation and the Jiufotang Formation. The Jehol biota is defined, spatially and temporally, either broadly [CHE 88] or in a much narrower way [PAN 13]. The fossiliferous sites have been explored and studied intensively for about 20 years. They are primarily known for their very well-preserved fossils of dinosaurs, birds, mammals and plants, but these fossiliferous levels have also been known for several decades for their plentiful fish fossils. The first bony fish to be described there was *Lycoptera* by Sauvage in 1880 [SAU 80].

Lycoptera was a basal osteoglossomorph characteristic of several Lower Cretaceous sites in northern China, eastern Mongolia, eastern Siberia and Korea. This fish gives its name to a large biogeographical group, the "*Lycoptera* fauna" [CHA 96, CHA 04]. This assemblage dates from the Barremian–Aptian. It can be distinguished from a younger Aptian–Albian assemblage located further south that contains the "*Mescoclupea* fauna", named after an ichthyodectiform that can be found there in large numbers. This fauna was initially identified in southern China [CHA 96] and its area was later extended to Japan and South Korea. It was then renamed "*Mesoclupea-Paraclupea* fauna" [CHA 04]. Contrary to more northern fauna, it does not contain lycopterids, but can be characterized by other basal osteoglossomorphs and a greater variety of sinamiids (halecomorphs). There are also two ginglymodians and of course, the ichthyodectiform *Mesoclupea*, whose general morphology is reminiscent of basal osteoglossomorphs.

When the Great Lakes of the *Lycoptera* fauna developed, the climate was hot-temperate to semi-arid with an annual temperature of about 15°C, measured with a geochemical carbonate analysis [AMI 15]. The region experienced substantial volcanic activity. The lakes were only about 10 m deep, as indicated by the abundance of mayfly larva. A recent taphonomic study of the *Lycoptera* specimens, sometimes preserved in the form of mass mortality events [PAN 15], makes it possible to reconstruct the following annual cycle: at the end of spring and the beginning of summer, when the lake water was well oxygenated and full of life, dead fish decomposed rapidly; during the summer, the amount of organic matter caused eutrophication in the water and made the bottom anoxic, bringing about the death and preservation of benthos (molluscs, crustaceans, mayfly nymphs, but few fishes); finally, in winter, the drop in temperature caused layers of water to mix, leading to mass mortality events for fishes that were then easily preserved on the bottom. The Jehol biota contains primarily *Lycoptera* and other basal osteoglossomorphs (cf. 4.14). The other clades present were rare ginglymodians (*Neolepidotes*), some sinamiids, various members of a family of acipenseriforms that is endemic to east Asia, peipiaosteids, a polyodontid (*Protopsephurus liui*) and a coccolepidid (*Coccolepis yumenensis*). The youngest and western-most component of *Lycoptera* fauna, located in Xinjiang, is placed within the Shenchinkow Formation by some authors [CHA 96, CHA 04]. This formation preserved a small radiation of basal teleosts, the siyuichthyids, of which there are five genera and five species described.

In 1994, Yabumoto [YAB 94] described the fauna of Wakino, a rich assemblage of freshwater fishes from Kyushu in southern Japan. Yabumoto recognized more than 20 species divided into three distinct assemblages. The most basal, named *Nipponamia-Aokiichthys*, contains a ginglymodian and an amiiform, an ichthyodectiform and a small radiation of species of the osteoglossomorph *Aokiichthys*. This genus was later attributed to *Paralycoptera* by Xu and Chang [XU 09]. The intermediary assemblage, named *Paraleptolepis-Wakinoichthys*, contains an indeterminate basal teleost, *Paraleptolepis*, an ichthyodectiform and the osteoglossomorph *Wakinoichthys*. The third and youngest assemblage also contains an ichthyodectiform and some osteoglossomorphs, as well as an abundance of clupeomorphs attributed with uncertainty to the genus *Diplomystus*. The most basal assemblage is considered to be equivalent to the "*Mesoclupea* fauna" in China and the two other assemblages are considered to be endemic [YAB 94]. The appearance of *Diplomystus*, which probably belongs to another genus of clupeoids according to Grande [GRA 82], is an interesting feature of this succession of assemblages. An ichthyofauna of equivalent age was studied from a subgroup in Nagdong in South Korea, notably including *Wakinoichthys*, *Lepidotes* and *Sinamia* [YAB 06], but the presence of an elopomorph in the latter assemblage indicated to these authors that the site was located closer to the sea than the Japanese assemblage.

A second site from in the Japanese Lower Cretaceous yielded a freshwater fish fauna from the subgroup Itoshiro, Tetori Group located in the center of the country. The fossiliferous levels date from the Berriasian-Hauterivian. Yabumoto [YAB 05] reported a *Lepidotes*, based on isolated scales, the sinamiid *Sinamia kukurihime*, an indeterminate pachycormid and the osteoglossiform *Tetoriichthys kuwajimaensis*.

2.3.3.2. South East Asia

The Cretaceous freshwater fishes of southeastern Asia are especially prominent in the continental series of the Khorat Plateau in northeast Thailand. The vertebrates of the Khorat Group, whose age extends from the Late Jurassic to the Aptian, have been studied since their discovery in the 1980s. The fishes were initially recognized in the form of isolated ganoid scales attributed to semionotids. The Phu Nam Jun site, located in the Phu Kradung Formation dating from the top of the Late Jurassic and the base of the Early Cretaceous, yielded several hundred specimens, often well preserved, belonging to a unique species of ginglymodian, *Thaiichthys buddhabutrensis* [CAV 03, CAV 13a]. In the same deposit, a specimen was attributed to a second taxon, *Isanichthys palustris* [CAV 06]. The taphonomy of the Phu Nam Jun site and the exclusive presence of fishes (no tetrapods have been discovered and the only other known vertebrate is a lungfish) indicate that the environment of the deposit was a dry pond at the bottom of which fish died and were buried in the sediment [DEE 09]. A second species of *Isanichthys*, *Isanichthys lertboosi* [DEE 14], was described at the Phu Noi site, which is probably slightly older than the Phu Nam Jun site. Contrary to Phu Nam Jun, the Phu Noi site contains a wide variety of terrestrial and aquatic vertebrates, which corresponds to an accumulation of carcasses in an environment where energy was stronger. Recently, a distinct ginglymodian, *Khoratichthys*, was recognized in the Phu Kradung Formation [DEE 16]. Recent phylogenetic analyses have reconstructed the relations within the old group of the "semionotiforms" [CAV 10, LÓP 12] and indicate that *Thaiichthys*, *Isanichthys* and possibly *Khoratichthys* are basal lepisosteiforms, a clade including modern gars (cf. 4.10.1). Several locations of the Phu Kradung Formation provided isolated dental plaques of lungfishes and the Phu Nam Jun site yielded a skull and a mandible attributed to *Ferganoceratodus martini* [CAV 07a]. As the generic name implies, the presence of this taxon indicates biogeographical relations with central Asia, and more precisely with the Fergana basin in Kyrgyzstan where the type species comes from. To my knowledge, the Thai lungfish represents the last occurrence of lungfish in Eurasia.

The other formations of the Khorat Group, especially the Sao Khua and Khok Kruat formations, have for the moment provided mostly isolated elements of fishes [CAV 09, CAV 14]. Based on the morphology and the microstructure of the scales, several taxa of ginglymodians (the clade that contains the ancient "semionotiforms") can be recognized. In the youngest formation in the group, the Khok Kruat Formation dating from the Aptian, the Phu Pok site produced several isolated skeletal elements and two partially articulated skulls of a holostean fish belonging to the sinamiids, a family endemic to Asia, *Siamamia naga* [CAV 07b].

2.3.3.3. India

The Lameta Formation in the Madhya Pradesh in India, dating from the Maastrichtian, contains an assemblage of fishes that seems to correspond to a mix of marine and freshwater forms. The fossils are isolated elements identifiable at the family level or in some cases at the generic level. There are ginglymodians, including a lepisosteid and *"Lepidotes"*, to which may be added some teeth attributed to "eotrigonodontids" (*Indotrigonodon, Stephanodus, Eotrigonodon*), some pycnodontids, an aspidorhynchid and a phareodontid [JAI 83, MOH 96]. Among the more marine elements, there are teeth attributed to aulopiforms.

2.3.4. Africa

2.3.4.1. Cocobeach and Rio Mundi

The Cocobeach sequence in Gabon and Rio Mundi in Equatorial Guinea corresponds to brackish facies dating from the Aptian. The fish fauna includes an indeterminate amiid, the aspidorhynchid *Vinctifer*, the ichthyodectiform *Chiromystus guinensis*, the chanids *Parachanos aethiopicus* and *Parachanos batai*, some indeterminate teleosts such as *Wenzichthys congolensis* and several ellimmichthyiforms including *Ellimmichthys goodi* and *Ellimma guineensis* [ARA 35, CAS 71, GAY 89].

2.3.4.2. The Continental Intercalary

In 1931, Kilian [KIL 31] named a sedimentary series of continental origin that was sandwiched between marine rocks from the Paleozoic below and marly limestones deposited during the Cenomanian marine incursion above. Because of the absence of stratigraphic fossils, the series is difficult to date. In certain regions like Niger, its base may go back to the Middle or Upper Jurassic [RAU 09], but in a large part of its extension, the Continental Intercalary is considered to date from the Early Cretaceous. At the beginning of the 20th Century in eastern North Africa, in Egypt, the German paleontologist Ernst Stromer discovered a rich continental fauna of reptiles and fishes in Bahariya, named after the nearby oasis. The group has been dated to the Cenomanian because of the presence of marine intercalations. At the western edge of North Africa, in Morocco, an assemblage of fishes and reptiles was announced in the mid-20th Century and was quickly compared to the Bahariya assemblage by Choubert *et al.* [CHO 52]. However, the age of these levels has been regularly considered to date from the Early Cretaceous, according to Lavocat [LAV 54] for example. A recent review [LE 12], based on the comparison of vertebrate assemblages in North Africa, made it possible to define a succession of levels of vertebrate assemblages in North Africa. The oldest level, dating from the Hauterivian–

Barremian, corresponds to assemblages from the Douiret Formation in Tunisia, the Cabao Formation in Libya, the El Rhaz Formation in Niger and the Koum basin in Cameroon. The next level, dating from the Aptian–Albian, corresponds to assemblages from Tilemsi in Mali, the Aïn el Guettar Formation in Tunisia and the Kiklah Formation in Libya. The fossiliferous part of the Moroccan Continental Intercalary, which is included in the Ifézouane Formation [ETT 04] and also known as the "Kem Kem beds" [SER 96], corresponds to the upper part of the series that can be dated, like the Bahariya fauna, to the Early Cenomanian [CAV 10].

The remains of pre-Cenomanian bony fishes from the Continental Intercalary were described by Tabaste [TAB 63] on the basis of fossils coming from various locations. She described various species of lungfish, a *Lepidotes*, a mawsoniid coelacanth, and *Stromerichthys* and *Eotrigonodon*, two genera whose affinities were unclear at the time (cf. 4.10.1). Mention should be made of the lepisosteiform *Pliodetes* from the Aptian of Gadoufaoua in Niger [WEN 99]. The Cenomanian ichthyological assemblage of Bahariya was studied by Weiler [WEI 35] and later reviewed by Schaal [SCH 84], among other authors. The Moroccan ichthyological fauna of the Ifézouane Formation, or fauna from the "Kem Kem beds", was recently reviewed by Cavin et al. [CAV 15]. This fauna, quite similar to that of Bahariya, contained the lungfish *Ceratodus humei*, *Arganodus tiguidiensis* and *"Neoceratodus" africanus*, the mawsoniid coelacanths *Mawsonia* and probably *Axelrodichthys*, some polypterids including *Bawitius*, the halecomorph *Calamopleurus africanus* and a rather large variety of ginglymodians, including *Adrianaichthys pankowskii*, *Obaichthys africanus*, *Dentilepisosteus kemkemensis* and *Oniichthys falipoui*. Some teleosts were also present, including the ichthyodectiform *Aidachar pankowskii*, probably several osteoglossomorphs, notably *Palaeonotopterus greenwoodi* and possibly *Erfoudichthys rosae*, as well as a tselfatiiform, *Concavotectum moroccensis*. The environment was probably a gigantic delta that covered a large part of North Africa. The fish fauna is characterized by an abundance of large species [CAV 15]. It was a food source for a large population of *Spinosaurus*, a semiaquatic theropod dinosaur [LÄN 13]. An assemblage of fishes similar to that of the Kem Kem beds, which also supported a large number of spinosaurs, was described in the Guir basin in Algeria [BEN 15].

Within the Kem Kem beds series is an assemblage of freshwater fishes at the Djebel Oum Tkout site [DUT 99a]. The very fragile specimens were preserved in sediment deposited on the bottom of a small lake inside a large delta. The fossils are very well preserved and some soft tissues, like muscle fibers, were sometimes preserved [DUT 99, GUE 14]. This ichthyofauna can be distinguished from the Kem Kem bed fauna. It contains the polypterid *Serenoichthys* [DUT 99b] and other cladistia that have not yet been described, a possible characiform that is still little known, a paraclupeid, an acanthomorph with unclear affinities, *Spinocaudichthys*, [FIL 01] and an actinopterygian with unclear relationships, *Diplospondichthys* [FIL 04]. *Spinocaudichthys* may be close to basal paracanthopterygians and *Diplospondichthys* may be related to *Tomognathus*, a very specialized marine halecomorph from the Early and "Middle" Cretaceous [FOR 06, CAV 12].

Among the other Cretaceous assemblages of freshwater fishes in North Africa is the Cenomanian site of Wadi Milk in Sudan. It contains various polypterids, which are represented by isolated spines and whose taxic diversity has probably been overestimated, some osteoglossomorphs and some possible characiforms [WER 94, GAY 02]. In the younger site of In Becetem in Niger (Coniacian–Santonian), a wide variety of polypterids, whose diversity is also questionable, was also described on the basis of isolated elements [GAY 02].

2.3.5. *South America*

2.3.5.1. *Early Cretaceous*

The Barremian Ilhas Formation in the Recôncavo basin in Bahia state, Brazil, contains an indeterminate ginglymodian, the ellimmichthyiforms *Scutatuspinosus itapagipensis* and *Ellimmichthys longicostatus*, the amiid *Calamopleurus mawsoni*, the ichthyodectiforms *Chiromystus mawsoni* and *Chiromystus woodwardi*, and *Scombroclupea scutata*, which replaces *Scombroclupea bahiaensis* according to Figueiredo [FIG 05]. The composition of this fauna and those that follow (Areado, Crato, Marizal) are related to the East African ichthyofauna in Cocobeach and Rio Mundi, which are approximately the same age. They are evidence of the development of rifts that caused the opening of the South Atlantic and illustrate a time when the proximity of the fauna from the two continents was still high [MAI 00].

The Areado Formation in Minas Gerais state in Brazil, which probably dates from the Aptian, yielded the chanids *Dastilbe moraesi* and *Laeliichthys ancestralis*. The latter was for a long time included in the lineage of the osteoglossomorph *Arapaima*, but it is more probably a stem osteoglossomorph [FOR 10].

The Crato Formation located in northeastern Brazil is a series of laminate limestone dating from the Aptian. It corresponds to a conservation Lagerstätte due to the excellent quality of conservation of its flora and fauna. In all likelihood, the deposit environment was a lagoon with variable salinity, but some fishes present should be highlighted because they belong to groups which are primarily freshwater elsewhere. The most abundant specimens belong to the chanid *Dastilbe crandalli*. The other taxa, represented by fewer specimens, include the ginglymodian *Araripelepidotes*, the amiid *Cratoamia gondwanica*, an ophiopsid, an aspidorhynchid, the ichthyodectiform *Cladocyclus gardneri*, an albuloid and a mawsoniid coelacanth [BRI 98, BRI 08].

The Marizal Formation in the Recôncavo basin in the Bahia state of Brazil, which is probably Aptian in age, corresponds to a fairly similar environment to the Crato Formation. It was a lagoon with variable salinity and episodic marine connections. The ichthyofauna was dominated by a teleost with unclear affinities, *Clupavus brasiliensis*, and by the euteleost *Britoichthys marizalensis*. It also contained the aspidorhynchiforms *Vinctifer longirostris* and *Vinctifer comptoni*, an ichthyodectiform, the ophiopsid *Placidichthys tucanensis* and the chanid *Nanaichthys longipinnus* [MAI 00, BRI 08, AMA 12].

The Muribeca Formation in the Sergipe-Alagoas basin contained an ichthyofauna comparable to the previous formations, namely the aspidorhynchid *V. comptoni*, the ichthyodectiform *Chiromystus mawsoni*, the chanid *Dastilbe crandalli* and the ellimmichthyiform *Ellimma branneri*.

In the Albian Lagarcito Formation in Argentina, the "Loma del *Pterodaustro*" site, where the filter-feeding pterosaur of the same name was found, also contained some fishes. The paleoenvironment was a large, shallow lake on an alluvial plain in a semiarid climate [CHI 98]. The ginglymodians from this site were initially attributed to three distinct genera of "semionotids", but they have since been classified within a single species,

Neosemionotus puntanus [LÓP 07]. A few indeterminate pleuropholids are also present.

The Albian formations of Codo and Santana contain rich assemblages of fishes, some taxa of which also occur in freshwater environments. However, the sedimentology suggests that the environments of these formations correspond mostly to closed marine areas, a hypothesis also supported by the presence of typically marine taxa (pachyrhizodontoids, araripichthyids, elopiforms). Consequently, these ichthyofaunas are not detailed here.

2.3.5.2. *Late Cretaceous*

The Albian-Cenomanian Alcântara Formation in northeastern Brazil contains an assemblage of vertebrates close to the Moroccan Kem Kem beds assemblage, which includes bony fishes. It yielded the remains of the polypterid *Bartschichthys*, pycnodontids, the ginglymodians "*Lepidotes*", *Eotrigonodon* and *Stephanodus*. The latter are considered to be indeterminate teleosts [CAN 11] but they may be referred to the ginglymodians. Also present were the coelacanth *Mawsonia* and some lungfish, including *Arganodus* and several species of *Ceratodus*. Candeiro *et al.* [CAN 11] notes that the assemblage of vertebrates in this formation is more similar to the "Middle" Cretaceous assemblages from North Africa (the Continental Intercalary) than to the assemblages in southern South America. In the Cenomanian, continental crossings were probably still possible both toward North Africa and toward southern South America. However, the latter were used less often than the former due to very different climatic and environmental conditions between the north and south of South America. This example illustrates the substantial influence of environmental factors over the simple physical opportunities via geographical connections in order to explain dispersal of freshwater fishes.

The Baurú Group in southeastern Brazil includes formations that yielded fossils of continental fishes. The most abundant is the Marília Formation dating from the Maastrichtian. It contains isolated elements of lungfish, an amiid, a lepisosteid named "*Lepisosteus*" *cominatoi*, although this is considered *nomen dubium* [GRA 10], as well as unidentified characiforms, siluriforms and perciforms [MAR 15].

The Maastrichtian Yacoraite Formation in northern Argentina was formed in a coastal environment containing a primary freshwater ichthyofauna represented by isolated fossils. It contains lepisosteiforms, pycnodontiforms, siluriforms and the gasteroclupeid *Gasteroclupea* [ARR 96].

Various deposits in Bolivia contain assemblages of freshwater fishes in two formations that follow one another and straddle the Cretaceous–Palaeogene boundary. The upper part of the El Molino Formation dates from the Maastrichtian and according to Gayet *et al.* [GAY 93], it corresponds to a mixed environment combining marine elements with continental environments (the base of the formation would be more clearly marine). The fossils are mainly isolated remains, but there are also fragments of more complete specimens, notably some braincases of siluriforms. According to Gayet and Meunier [GAY 98], and considering the systematic revisions posterior to the initial descriptions, the list of taxa includes the polypteriform *Latinopollia suarezi* and the polypterid *Dagetella sudamericana*, several ginglymodians including a possible lepisosteid and a peculiar form, *Lepidotyle enigmatica*, which is characterized by a remodeling of the ganoin layer covering the scales [GAY 92]. The teeth of *"Stephanodus"* could possibly be attributed to one of the ginglymodians. The percichthyid *Percichthys hondoensis* was also recorded as well as the remains of pycnodontids, osteoglossomorphs (*Brychaetus* and *Phareodusichthys*) and gasteroclupeids (*Gasteroclupea*). Characiforms such as *Tiupampichthys intermedius* and other forms with unclear affinities were also noted, as well as a siluriform belonging to the extinct andinichthyids, *Andinichthys*, and a possible diplomystid.

2.3.6. *North America*

Grande and Grande [GRA 99] considered the fossil record of Cretaceous freshwater fishes in North America to be poor. However, despite the rarity of occurrences, these have a great significance for our understanding of the evolutionary history of several freshwater lineages on a global scale. Indeed, fluvial deposits from the Cretaceous in North America illustrate a transitional period for freshwater fauna that corresponds to a shift from assemblages composed of taxa (at the level of families or suborders) on the point of extinction and new families that are still represented in nature today [NEU 05].

2.3.6.1. *Milk River, Oldman, Dinosaur Park, Horseshoe Canyon and Scollard formations (Canada)*

The Milk River Formation in Alberta dates from the Santonian–Campanian boundary. The marine influences are stronger than in the formations that follow after, but it, nonetheless, contains continental vertebrates, including a small assemblage of freshwater fishes. It includes the oldest known pike (esocid), *Estesesox* [WIL 92] and the clupeomorph ellimmichthyiform *Horseshoeichthys* [NEW 10].

The Oldman Formation from the Campanian in the south of Alberta was formed in a flood plain environment. It contains a rich fauna of dinosaurs. The fishes found there belong partly to ancient lineages and partly to families that appear in the fossil record for the first time there. Some of the ancient families rapidly went extinct, like the aspidorhynchid *Belonostomus*, and others survived to the present day, like the sturgeon *Acipenser* and the gar *Lepisosteus*. In the new families, there are the most ancient representatives of modern lineages such as pikes (esocids) already represented by two genera, *Estesesox* and *Oldmanesox* [WIL 92]. With them are two representatives of probable freshwater elopomorphs, *Paratarpon* [BAR 70] and the enigmatic phyllodontid *Paralbula*, as well as the osteoglossomorph *Cretophareodus* [LI 96].

The Campanian Dinosaur Park Formation rests on the Oldman Formation in Alberta. It was deposited in an environment of alluvial and coastal plains. As its name indicates, it contains various dinosaurs, but also the remains of fishes. These are generally disarticulated because the areas in which they are found are of fluvial origin with a relatively high energy. This can make the identification of taxa difficult. The ichthyofauna of this formation is similar to that of the Oldman Formation [NEU 05]. It includes *Belonostomus*, acipenseriforms represented by taxa similar to the sturgeon (*Acipenser*), paddlefish (*Polyodon*), gars (ginglymodians), and amiids (halecomorphs). Two others, which are called holosteans A and B and are known exclusively from isolated remains, can be added to this list [BRI 90]. Among the teleosts in the Dinosaur Park Formation are *Paratarpon* [BAR 70], an elopomorph close to the modern tarpon (*Megalops*), which is marine, and *Cretophareodus*, an osteoglossomorph [LI 96]. *Paralbula* is also present along with *Coriops*, another phyllodontid from the Upper Cretaceous in the United States. Neuman and Brinkman [NEU 05] do not exclude the possibility that *Coriops* had affinities with the osteoglossomorphs. The two esociforms *Estesesox* and *Oldmanesox* are present, as well as at least eight other small teleosts, including several acanthomorphs. The large proportion of teleosts, including the acanthomorphs, is a surprising feature of the Dinosaur Park Formation assemblage highlighted by Neuman and Brinkman [NUE 05]. Recently, the remains of characiforms were also described from this formation [NEW 09].

The Horseshoe Canyon Formation in Alberta, dating from the Campanian–Maastrichtian boundary, corresponds to various environments including flood plains, estuary channels and swamps. The clupeomorph ellimmichthyiform *Horseshoeichthys* [NEW 10] comes from this formation. The Scollard Formation near Rumsey in Alberta, dating from the Cretaceous–Paleogene boundary (between 66 and 63 million years ago) contained the amiid *Cyclurus fragosus* [GRA 98].

2.3.6.2. The Lance and Hell Creek formations (USA)

The Maastrichtian formations of Lance in Wyoming, and Hell Creek in Montana, North Dakota, South Dakota and Wyoming, have long been studied. They contain very rich assemblages of terrestrial vertebrates and more modest freshwater fish fauna. The Hell Creek Formation fauna is slightly more diverse than that of the Lance Formation, so we will detail that formation here, knowing that all of the elements in the Lance Formation are present in the Hell Creek Formation [BRI 14]. The families are similar to those in the Canadian formations. There are indeterminate species of modern genera of gars [GRA 10] as well as a "semionotid" ginglymodians, two extinct genera of amiids, *Cyclurus fragosus* and *Melvius thomasi*, the aspidorhynchid *Belonostomus longirostris*, the acipenserid *Acipenser* and *Protoscaphirhynchus squamosus*, the polyodontid *Paleopsephurus wilsoni*, the albulid or osteoglossomorph *Coriops amnicolus*, two indeterminate hiodontids and the phyllodontid *Phyllodus paukatoi* [EST 69, EST 78]. In addition, an indeterminate ostariophysian that may be attributed to a siluriform [BRI 14], the esocid *Estesesox foxi* [WIL 92], the indeterminate percopsiform *Priscacara* and the indeterminate acanthomorph *Platacodon nanus* were also described.

2.3.7. Madagascar

The Ankazomihaboka Series, dating from the Coniacian–Santonian in the Mahajanga basin in northeastern Madagascar, yielded a mawsoniid coelacanth [GOT 04] and the amiid *Melvius* [GOT 09]. Above, the Maevarano Formation from the Upper Campanian–Maastrichtian also yielded an assemblage of freshwater fishes in the form of isolated elements. It includes the remains of pycnodontids, *Enchodus* (a genus generally found in marine deposits); indeterminate siluriforms, phyllodontids, and albulids; possible characiforms and cypriniforms (these identifications need to be confirmed) and a possible perciform [OST 12].

Assemblages of Freshwater Fishes in the Cenozoic

3.1. Paleogene

During the Paleogene, from 66 to 23 million years ago, the face of the Earth began to resemble what we know it as today. However, India had only just begun to collide with Eurasia, South America was separated from North America, and Europe, where the Alps have not yet uplifted, was separated from Asia by the Turgai Strait. The climate was still hotter than it is today, especially during an extreme heat event at the Paleocene-Eocene boundary. The approximate geographic location of the main sites containing freshwater fishes in the Paleogene is presented in Figure 3.1.

Palaeocene (35 myrs)

Figure 3.1. *Main sites and formations from the Paleogene that yielded faunas of freshwater fishes positioned on a paleogeographic map of the Eocene. For a color version of the figure, see www.iste.co.uk/cavin/fishes.zip*

3.1.1. *North America*

3.1.1.1. *Paleocene*

The Paskapoo Formation in Alberta, Canada, and the Tullock Formation in Montana, USA, correspond to environments of rivers and swamps dating from the Late Paleocene (Thanetian). The freshwater fishes that they contain are in part similar to fishes in the same regions from the Maastrichtian. These include acipenseriforms, lepisosteids and amiids, as well as phyllodontids, osteoglossomorphs (hiodontids and osteoglossoids) and esocids, although aspidorhynchids are absent. We also find an indeterminate gonorynchiform (Paskapoo) and one of the most ancient percopsiforms *Massamorichthys wilsoni* [MUR 96]. In the Paskapoo Formation, there was also an osmerid, *Speirsaenigma lindoei* [WIL 91], the most ancient known catostomid, and an indeterminate neoteleost attributed to the asineopids. One level in the lower part of the Paskapoo Formation yielded a small area containing a large number of articulated specimens preserved in a fluvial environment. The dominant genus were the percopsid *Massamorichthys* followed by, in fewer number, the osmerid *Speirsaenigma* and rare specimens of the osteoglossomorph *Joffrichthys*. This type of preservation, which is rare in a fluvial environment, is interpreted as the result of a mass mortality event following the drying up of a spawning ground such as an abandoned channel, a pond or a shallow lake [WIL 96].

Among the other freshwater fishes from the Paleocene of North America are the siluriforms *Rhineastes* and *Astephus*, belonging to the ariids and ictalurids, respectively, a palaeolabrid, and *Acipenser* [CAV 98].

3.1.1.2. *Eocene*

Recently, Divay and Murray [DIV 16] described a small ichthyofauna from the Early Eocene in the Wasatch Formation in Wyoming. This formation corresponds to a shallow river environment with a weak current that preceded the lacustrine environments characteristic of the Green River Formation. The taxa described were preserved as disarticulated elements and correspond, at the generic level, to the taxa of the Green River lakes: a lepisosteid and an amiid, the ellimmichthyiform *Diplomystus*, the gonorynchid *Notogoneus*, a probable amblyopsid, some centrarchids and *"Priscacara"*. Divay and Murray [DIV 16] highlight the similarities between this fluvial fauna and the slightly younger lacustrine fauna of Green River, despite a significant warming of the climate during the interval separating these two faunas. They also note that at the family level, this fauna is rather similar to the fauna of the Late Cretaceous in North America.

The Green River Formation corresponds to a deposit accumulated at the bottom of intramontane lakes that are now exposed in the states of Colorado, Wyoming and Utah. It is probably one of the best studied and richest lacustrine accumulations in the world for the Cenozoic era. The corresponding site in the Mesozoic would be the Jehol biota in China. Grande [GRA 84] published a monograph that presents the paleoenvironment and the fauna of this formation, with a particular focus on the fishes. The fossils are mostly concentrated in a series that corresponds to a period of 5 million years at the end of Early Eocene, between 53.5 and 48.5 million years ago. The environment at the time that the sediments were deposited corresponds to three distinct lakes; Fossil Lake, Lake Gosiute and Lake Uinta, which have slightly different fish fauna. Lake Uinta remained for a longer period than the two other lakes, from the Late Paleocene to the Late Eocene. In this area, there are the polyodontid *Crossopholis magnicaudatus*, various holosteans including lepisosteids and amiids, the hiodontid *Hiodon falcatus* [HIL 08], two osteoglossomorphs, *Phareodus encaustus* and *P. testis*, at least two species of clupeiforms, *Knightia humilis* and *Knightia alta*, an ellimmichthyiform, *Diplomystus dentatus*, a gonorynchid, *Notogoneus osculus*, the catostomid cypriniform *Amyzon*, two siluriforms, *Astephus antiquus* and *Hypsidoris farsonensis*, two percopsids, *Amphiplaga* and *Erismatopterus*, an acanthomorph with unclear affinities, *Asineops squamifrons*, and some percoids, *Mioplosus labracoides*, *Priscacara serata* and *P. liops*. The relative wide diversity of holosteans is remarkable. The Fossil Butte Monument from the Eocene contains four species of lepisosteids, *Lepisosteus bemisi*, *Atractosteus simplex*, *Atractosteus atrox* and *Masillosteus janeae*, as well as two species of amiids, *Amia pattersoni* and *Cyclurus gurleyi*, to which can be added the lepisosteid *Cuneatus cuneatus* in the Upper Paleocene or lower Eocene layers of Lake Uinta [GRA 98, GRA 10].

The Clarno Formation in Oregon produced a fauna from the middle Eocene that contains families that are characteristic of freshwater environments in North America during the Cenozoic: amiids, hiodontids, catostomids and siluriforms [CAV 98]. It is the same for the Florissant Formation in Colorado that contains *Amia scutata*, catostomids, ictalurids and percopsiforms [DIV 15, GRA 98] and the Late Eocene–Early Oligocene

John Day Formation in Oregon, which, along with the Cypress Hills Formation, contains the most ancient North American cypriniform.

Contemporary levels of the Green River Formation are known in British Columbia, Canada. They are particularly interesting because they correspond to a more northern fauna than in the Green River Formation. Several sites that correspond to deposits of lakes and rivers have yielded a rich assemblage of fishes that include taxa similar to those in Green River at the familial and generic levels. These include *"Amia" hesperia*, a hiodontid, *Hiodon rosei*, the catostomids *Amyzon mentale* and *Amyzon aggregatum*, a percopsid, *Libotonius blakeburnensis*, and a priscacarid, *Priscacara aquilonia* [WIL 77]. The most ancient representative of the salmonids, *Eosalmo driftwoodensis*, unknown in the deposit stateside, was also described there. Lacustrine varves corresponding to annual cycles were preserved in the horsefly site. These make it possible to demonstrate that some events were seasonal, such as fish dying from hunger or lack of oxygen in winter [WIL 77].

The Cypress Hills Formation in Saskatchewan, Canada, which dates from the Late Eocene or Early Oligocene, recently yielded an interesting fauna composed of isolated elements that were deposited in a fluvial environment [DIV 15]. The fauna includes "ancient" elements such as the lepisosteid *Lepisosteus*, an amiin and a hiodontid. Among the more recently arrived taxa to the North American fauna are a catostomid (cypriniform) and the ictalurid *Astephus* (siluriform), salmoniforms and a perciform with unclear affinities, *Mioplosus*. Most importantly, this deposit includes one of the oldest occurrences of a leuciscin perciform, aff. *Ptychocheilus*, as well as possible amblyopsids, moronids and centrarchids. This ichthyofauna offers a glimpse into the modern ichthyofauna.

3.1.2. *Europe*

The Paleogene fauna of Europe are generally considered to be isolated from the fauna of neighbouring continents, especially North America [CAV 98].

3.1.2.1. *Ménat and Cernay*

The site of Ménat, in Puy-de-Dôme in France, represents an environment that is quite similar to Messel in Germany, but it is slightly older (Late

Paleocene, Thanetian). It formed in a lake at the bottom of a maar that appeared during the first phases of volcanic activity in the Massif Central. The fossils were preserved in bituminous diatomites. The ichthyofauna is not very diverse. It includes the amiid *Cyclurus valenciennesi*, a teleost with unclear affinities, *Thaumaturus brongniarti*, and the percichthyid *Properca angusta* [GAU 79a].

The Late Paleocene site of Cernay, near Reims, France, yielded the remains of *Amia* [GRA 98].

3.1.2.2. *Messel*

The middle Eocene site of Messel was also formed in a tropical lake that was probably completely isolated during a large part of its existence in this geologically active region. The bottom of the lake, in a highly stratified water column, was anoxic. Outgassing was probably responsible for the death of terrestrial and aerial wildlife in the surrounding area (terrestrial mammals, bats, birds). The ichthyofauna of Messel is clearly dominated by the gars *Atractosteus messelensis* and *Cyclurus kehreri*, two genera that are currently able to survive in water with very low levels of dissolved oxygen. These taxa were native, unlike other rarer species that entered the lake during short periods when the lake was connected to surrounding rivers. Among the non-native elements is the eel, *Anguilla ignota*, known from only one specimen. It must have entered the lake through a river system feeding directly into the ocean since this species was probably catadromous, like those of the genus *Anguilla* today [MIC 12]. There is also a second genus of gar, *Masillosteus kelleri*, which has its sister species in the Green River Formation. The fossils of this fish are rare and probably indicate a non-native species. The other taxa present include a teleost with unclear affinities (*Thaumaturus*), a percichthyid (*Amphiperca*) and two moronids (*Palaeoperca* and *Rhenanoperca*) [MOR 04].

3.1.2.3. *Area around Paris*

The Montmartre gypsum and the supragypseous marls were deposited in a slightly brackish environment connected to the sea in the Late Eocene. Studied since the end of the 18th Century, the ichthyological assemblage includes some strictly freshwater fishes, like *Cyclurus ignotus*, and a teleost with unclear affinities, *Thaumaturus*. It also includes some fishes that are typical of environments with variable salinity such as an ambassid, *Dapalis*, and a gonorynchid, *Notogoneus*, as well as some euryhaline fishes such as the sturgeon, *Acipenser*, and the seabream, *Sparus* [GAU 81b].

3.1.2.4. Kutschlin

The middle or Late Eocene deposit of Kutschlin, in the Czech Republic, yielded the amiin *Cyclurus macrocephalus* [GRA 98].

3.1.2.5. Provence

In the Late Oligocene, large lakes developed in the Aix-en-Provence region of France, at the bottom of which large deposits of gypsum accumulated. Fish fossils from gypsum quarries have been studied for two centuries, but new fossiliferous outcrops have been discovered since. A small, systematic excavation in one of these outcrops made it possible to define the composition of the ichthyofauna and the deposit conditions of the gypsiferous marls [GAU 78b]. The ichthyofauna includes rare amiids, *Amia*, and the anguillid *Anguilla*, probably two species of the cyprinodontiform *Prolebias* [GAU 81a], an ambassid, *Dapalis*, a latid, *Eolates*, a gobiid, *Gobius*, a gonorynchid, *Notogoneus*, a mugilid, *Mugil*, a possible moronid, *Beaumontoperca beaumonti*, and a rare cyprinid, *Palaeotinca*. More recently, the gerreid *Pharisatichthys* was described from these levels [GAU 15]. The species from this family alive today are extremely tolerant of hypo- and hyperhaline conditions. *Amia* can be found in other sites in the same formation. The taxonomic composition and the small number of species in most of the banks indicate an environment with variable salinity. In particular, the presence of fry from the genus *Mugil* is an indication of possibly sporadic contact with the sea. This cannot be considered a typical freshwater assemblage.

Levels in the Apt basin yielded freshwater fishes like *Dapalis* and the oldest European pike, *Esox primaevus* [GAU 78a]. These layers are slightly older than those in the Aix basin, from the Early Oligocene.

Discoveries in the Oligocene of the Vaucluse and Hautes-Alpes Departments show an osmerid, *Enoplophthalmus*, a family that is typical in high latitudes today, coexisting with an ambassid, *Dapalis*, a family that is now tropical. These are indications that the seasons were not very distinct during this period ("seasonal equability"), according to Gaudant [GAU 13].

3.1.2.6. *Sieblos*

The Early Oligocene deposit of Sieblos in Hesse, Germany, yielded the amiin *Cyclurus oligocenicus*, an umbrid and some possible percichthyids [GRA 98].

3.1.3. *Asia*

3.1.3.1. *China, Mongolia and Central Asia*

The Naran Bulak Formation in the Nemegetin basin in Outer Mongolia corresponds to a lacustrine deposit that yielded the amiin *Cyclurus efremovi* as well as the remains of hiodontids [GRA 98]. In the province of Jilin, in northeastern China, the site of Huadian from the middle Eocene yielded a small assemblage containing the amiin *Cyclurus*, the catostomid *Plesiomyxocyprinus* and an indeterminate percoid [LIU 09, GAU 12]. In the province of Hunan, the site of Xiawanpu from the Early or Middle Eocene yielded the amiin *Cyclurus*, the catostomid *Amyzon*, the siluriform *Aoria* and two species of percoids from the genus *Tungtingichthys* [CHA 10]. The flora associated with this site indicates a hot and arid environment.

3.1.3.2. *India and Pakistan*

In India, the Paleocene Formation of Palana in Rajasthan yielded a fragment of a lepisosteid and the osteoglossiform *Taverneichthys bikanericus* [KUM 05]. The latter was re-examined by Taverne *et al.* [TAV 09]. The paleoenvironment of the lignite area of the Palana Formation, where the fish fossils are from, corresponds to the mouth of a river or a delta [KUM 05].

In Pakistan, the Kuldana Formation contains varied ichthyofauna from the lower or middle Eocene that was studied by Gayet [GAY 87] and Roe [ROE 91] at Chorlakki, and by Murray and Thewissen [MUR 08] in the regions of Gandas Kas, Jahalar and Thatta. This formation, where the primitive whale *Pakicetus* was found, corresponds to a fluvial environment composed of shallow river channels in an arid climate. The ichthyofauna includes a lepisosteid, an amiid, a possible cyprinid and a probable bagrid, a cyprinodontid, an indeterminate perciform and the channid *Anchichanna*, known only by isolated braincases. This fauna dates from one of the phases of collision between India and Eurasia, but it is likely that passage between the two continental masses was already well established by this time. Murray and Thewiessen [MUR 08] note similarities between the Pakistani fauna and African fauna from the Late Eocene–Early Oligocene site of Jebel Qatrani in Egypt, especially with the presences of channids, amiids and osteoglossids.

3.1.4. Africa and the Arabian Peninsula

3.1.4.1. Mahenge

This site from the middle Eocene (Lutetian) in Tanzania is noteworthy because of the environment that it represents and the fauna that it contains. The lake was formed in one of 50 kimberlite diatremes that developed between the Late Cretaceous and the start of the Cenozoic in the region of Singida in Tanzania. The lake of Mahenge measured about 370 m in diameter and was only about 20 m deep. The ichthyofauna, which already shows modern characteristics, contains a clupeomorph, *Palaeodenticeps tanganyikae*, five species of cichlids from the genus *Mahengechromis*, two genera and species of osteoglossomorphs, *Singida jacksonoides* and *Chauliopareion mahengeense*, a citharinoid, *Eocitharinus macrognathus*, a species of claroteid *Chrysichthys*, *Chrysichthys mahengeensis*, and the oldest known fossil knerrid, *Mahengichthys singidaensis* [MUR 01, MUR 03b, MUR 05, MUR 03, DAV 13].

3.1.4.2. North Africa

The Bartonian deposit of Dur At-Talah in Libya yielded mainly freshwater fauna even though the Idam Unit where it comes from was bioturbated and contains some elements from a marine or brackish environment [OTE 15]. The assemblage, primarily represented by isolated teeth, contains a lungfish, *Protopterus*, a polypterid, *Polypterus*, a phyllodontid, *Egertonia*, a gymnarchid, *Gymnarchus*, several siluriforms including *Chrysichthys* and a mochokid, several characiforms including *Hydrocynus* and other alestids, a channid, *Parachanna* and at least one cichlid. Some of these genera, including *Gymnarchus* and the goliath tigerfish (*Hydrocynus*), are, together with the discoveries at Fayoum, the oldest occurrences of these genera endemic to Africa.

Fayoum, in Egypt, includes several sites located in two Eocene formations, Qasr el Sagha and Birket Qarun, and in a formation dating from the Eocene-Oligocene boundary, Jebel Qatrani. The Birket Qarun Formation, which presents marine influences that may correspond to a delta, yielded an assemblage that was initially described by Ernst Stromer and was very similar to the assemblage from Dur At-Talah in Libya. In this assemblage, there are *Protopterus*, *Polypterus*, a pycnodontid, *Gymnarchus*, a mochokid, *Hydrocynus*, *Parachanna* [STR 05, MUR 10] and, additionally, a percoid with unclear affinities, *Weilerichthys fajumensis* [OTE 99]. The Jebel Qatrani Formation, deposited in a fluvial environment of a tropical forest in a monsoon climate, yielded the remains of lungfish, cichlids including *Tylochromis*, siluriforms, alestid characiforms, latids, channids and a clupeid, *Chasmoclupea* [MUR 02, MUR 04, MUR 05].

3.1.4.3. Tamaguélelt

The phosphate deposits in Tamaguélelt in Mali, dating from the Lutetian, correspond to a brackish environment according to Patterson and Longbottom [PAT 89] and a freshwater environment according to Grande and Bemis [GRA 98]. The area yielded the remains of lungfish, three species of pycnodontids, *Pycnodus maliensis*, *P. zeaformis* and *P. jonesae*, the claroteid siluriform *Nigerium wurnoense* [LON 10], osteoglossiforms, centropomids (or latids) and the amiid *Maliamia gigas* [PAT 89], a genus placed in the Calamopleurini tribe by Grande and Bemis [GRA 98].

3.1.4.4. Nsungwe Formation

The Nsungwe Formation in the Rukwa Rift basin in southeastern Tanzania dates from the Oligocene. It recently yielded a small fauna of alestid characiforms. The genus *Hydrocynus* and an indeterminate genus were found in this environment, which was shallowly fluvial and lacustrine [STE 16].

3.1.5. South America

3.1.5.1. Bolivia

The Santa Lucia Formation in Bolivia sits atop the El Molino Formation. It dates from the Danian and was deposited in environments that ranged from fluvial to lacustrine. The ichthyofauna resembles that of the El Molino Formation, notably with the presence of a lungfish, "*Ceratodus*" sp. [SCH 91], two polypterids, osteoglossomorphs (the genus *Phareodusichthys* and fragments of heterotins similar to *Arapaima*), gasterosteids (*Gasteroclupea*) and characiforms (including *Tiupampichthys*). However, the variety of siluriforms is greater here, with three distinct genera of andinichthyids, *Andinichthys bolivianenesis, Hoffstetterichthys pucai* and *Incaichthys suarezi*. It is interesting to note that the composition of the Paleocene ichthyofauna has some identical genera and does not differ substantially from the Maastrichtian ichthyofauna, although the site was not very far from the point of impact of the Chicxulub meteorite (cf. 5.3.2).

3.1.5.2. Argentina

The Laguna del Hunco Formation in Chubut, Argentina, is a site that dates from the Early Eocene. It corresponds to a lake in the hot climate of the Eocene climactic optimum in a volcanic region. Outgassing could be the cause of local mass extinctions. The site yielded an abundant flora, insects, amphibians and the siluriform *Bachmannia chubutensis* [AZP 11].

The Cañadón Hondo Formation near the San Jorge Gulf on the Atlantic coast is estimated to date from the Eocene. It yielded the percichthyid *Percichthys hondoensis* [ARR 96].

3.1.5.3. Peru

The Cenozoic deposits of Quebrada Cachiyacu near Contamana in Peru yielded various assemblages made up of microremains in about 20 distinct levels ranging from the Paleocene to the Late Miocene. These assemblages were the subject of a recent review [ANT 16]. Regarding freshwater fishes in the Paleogene, the Pozo Formation from the middle Eocene yielded characiforms belonging to the anostomids (cf. *Leporinus*), serrasalmids and cynodontids (*Hydrolycus*) in sites corresponding to a fluvial environment. The sites corresponding to alluvial plain deposits yielded loricariids and indeterminate characiforms. The Late Oligocene Chambira Formation yielded the same families of characiforms as in the previous formation plus erythrinids, loricariid and pimelodid siluriforms, and cichlids in sites representing small lakes.

3.1.6. Australia

The Redbank Plains Formation in Queensland dates from the Eocene. The remains of vertebrates are preserved in nodules that formed in a permanent lake. The ichthyofauna contains the lungfish *Neoceratodus gregori*, the osteoglossomorph *Phareodus queenslandicus*, the gonorynchiform

Notogoneus parvus and the percichthyid *Macquaria antiquus* [VIC 96, UMA 01].

3.2. Neogene

As mentioned in the introduction, the fossil record of the Neogene will not be described in this book for two reasons. First, the number of localities that date from between the Miocene and the Pleistocene is too great on a global scale to be described. Second, and most importantly, the majority of modern freshwater ichthyological faunas have existed since the Paleogene. Of course, several transformations occurred during the Neogene in continental environments subject to a global drop in temperature and the aridification of large regions but, at the intercontinental paleogeographic level, the families and often genera of today were already present. This does not prevent the diversification of actinopterygian families in continental environments from continuing in the Neogene as is indicated by an assessment of the evolution of phylogenetic diversification (cf. 5.2.1).

Several sites from the European Neogene have been studied, notably by Jean Gaudant and Bettina Reichenbacher. For African sites, some of the recent publications include [STE 01, OTE 10] and [OTE 09], and for sites in the Arabian Peninsula [OTE 01]. The North American Neogene sites containing the remains of freshwater fishes are indexed by Smith [SMI 81]. For South America, information can be found in [ARR 96, LUN 98, LUN 98] and [ANT 15]. For New Zealand, consult [MCD 10]. For central and eastern Asia, the Neogene sites that contain the remains of freshwater fishes (especially cypriniforms) are indexed by Chang and Chen [CHA 08].

4

Evolutionary Histories
of Freshwater Fishes

The clades presented in this chapter are organized according to the recent classification of Betancur *et al.* [BET 14] based on a molecular phylogenetic analysis of the modern families, with some modifications mentioned in the text. The extinct families were placed in this phylogeny based on the works of various authors that are not all mentioned here (see Figure 4.1).

4.1. Coelacanths (Actinistia)

Latimeria, the only living genus of coelacanth, was described by Smith in 1938 [SMI 38], 102 years after the description of the first fossil of *Coelacanthus* by Louis Agassiz in 1836. The discovery of the first specimen of *Latimeria* by Marjorie Courtenay-Latimer and its inclusion in a group of fishes that was believed to have been extinct since the end of the Mesozoic era had a huge impact on biologists and the public. The animal was considered to be a kind of intermediary between true bony fishes (actinopterygians) and vertebrates with legs, tetrapods. In coelacanths, the limbs are connected to the girdle by a single point, as in tetrapods, and the skeletal structure of their limbs resembles legs more than the rayed fins of the actinopterygians.

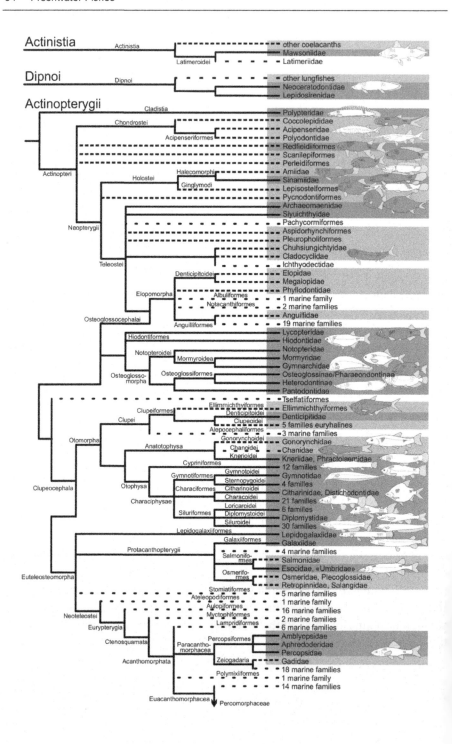

Actinistia

Dipnoi

Actinopterygii

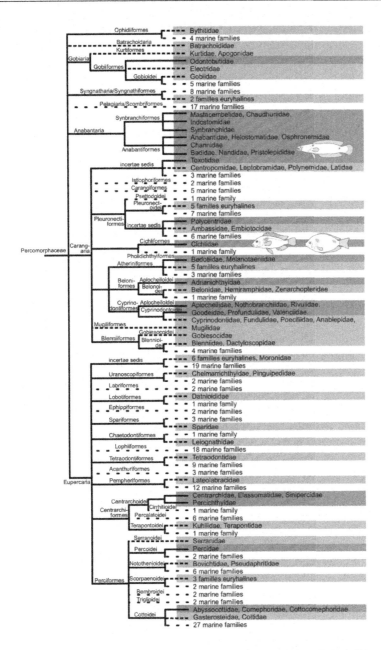

Figure 4.1. *Phylogenies of coelacanths (Actinistia), lungfishes (Dipnoi) and actinopterygians (Actinopterygii) from the Mesozoic and the Cenozoic. Dark green: freshwater taxa; light green: euryhaline taxa; gray: extinct taxa. For a color version of the figure, see www.iste.co.uk/cavin/fishes.zip*

The oldest known coelacanth fossils date from the Early Devonian. In the Devonian, coelacanths displayed a certain morphological disparity [SCH 56, CLO 91, FOR 98, SCH 04, FRI 06], particularly concerning fin placement, as illustrated by the genera *Holopterigius* and *Miguashaia*. All taxa from this period are found in fossiliferous deposits of marine origin. During the Late Paleozoic, the majority of species were always present in marine settings, with some exceptions in the Carboniferous and the Permian. The period of the greatest taxic diversity of coelacanths was the Triassic [CLO 91, FOR 98, SCH 04, CAV 13c], but it remains modest in comparison to other groups of fishes. The majority of taxa are always found in marine environments. The morphological diversity of coelacanths decreased after the Triassic and most species presented a generally uniform morphology that already largely resembled that of *Latimeria*. Their position in the evolutionary tree of vertebrates, which is supposedly as an intermediary between fishes and tetrapods, as well as their slow morphological evolution compared to other lineages earned them the epithet of "living fossil", an expression that remains up for debate (cf. [CAS 13] and [CAV 14]). From the Jurassic on, only two families are present in the fossil record: latimeriids and mawsoniids. Latimeriids are exclusively marine while mawsoniids represent the longest evolutionary experience of coelacanths in freshwater. Starting from the Triassic, mawsoniids have been essentially continental with *Chinlea* in the Chinle Formation and *Diplurus* in the Newark Formation (Figure 4.2). The only exception is the large marine genus from the Jurassic, *Trachymetopon*, which was present in Europe since the Toarcian of Holzmaden, Germany [DUT 15] and survived until the end of the Jurassic of the south of England and in Normandy [DUT 14]. In the Cretaceous, two genera cover almost the entire period in continental and brackish environments. The last known representatives of mawsoniids date from the Early Maastrichtian and it is likely that the lineage died out in the mass extinction at the Cretaceous–Paleogene boundary.

The first freshwater mawsoniid was discovered in the Bahia region of Brazil by Joseph Mawson, a railway engineer [MAI 16]. Initially believed to belong to a pterosaur and a plesiosaur, these fossils were later attributed to a coelacanth named *Mawsonia gigas* by Arthur Smith Woodward [MAW 07]. As the specific name indicates, the coelacanth in question was very large. One feature of this species, also found in all members of the family, is a very pronounced ornamentation of the skull bones formed by small crests forming a network pattern. Other species of *Mawsonia* were later described in Brazil, including *Mawsonia minor* by Woodward [WOO 08] and *Mawsonia brasiliensis* by Yabumoto [YAB 08], and in Africa, including *Mawsonia libyca* in Egypt by Weiler [WEI 35], *Mawsonia lavocati* in Morocco by Tabaste [TAB 63] and *Mawsonia tegamensis* in Niger by Wenz [WEN 75]. Recently, Carvalho and Maisey [CAR 08] studied a population of *Mawsonia gigas* from the Sanfranciscana basin in Minas Gerais in Brazil. Although primarily made up of isolated elements, this sample shows that the morphological variability was high within this species. According to the authors, it is likely that the various South American species mentioned above could all be related back to the type species, *M. gigas*, whereas the African species should be considered mawsoniids *incertae sedis*. In 1986, Maisey [MAI 86] described another genus of mawsoniid, *Axelrodichthys*, from the Santana Formation in northeastern Brazil. This formation, which also contains some *Mawsonia*, probably corresponds to a closed marine environment. In 2004, two works demonstrated that *Axelrodichthys* is probably also present in the Aptian of Ingal in Niger, the Santonian of Madagascar [GOT 04] and the Cenomanian of Morocco [CAV 04]. Finally, a recent interpretation of material from the Cenomanian of Morocco attributed to *M. lavocati* shows that it is more likely a species of *Axelrodichthys* [CAV 15]. It therefore appears that the two genera of mawsoniids from the Cretaceous, *Mawsonia* and *Axelrodichthys*, experienced vicariance phenomenon between Africa and South America after the South Atlantic opened (Figure 4.2). However, a systematic revision of the family is necessary to identify the sister species created by these vicariances. After the Cenomanian, the mawsoniids seem to be absent from the South American and African continents but can be found in the terminal Cretaceous in Europe. First, an isolated element, an angular bone, was found in the Maastrichtian of the Cruzy area in the south of France [CAV 05]. Then, a partial skull was discovered in another area in the south of France, Ventabren, which made it possible to define a new species attributed to the genus *Axelrodichthys*, *Axelrodichthys megadromos* [CAV 16]. Other remains have been recognized in various sites in the south of France and are waiting to be described.

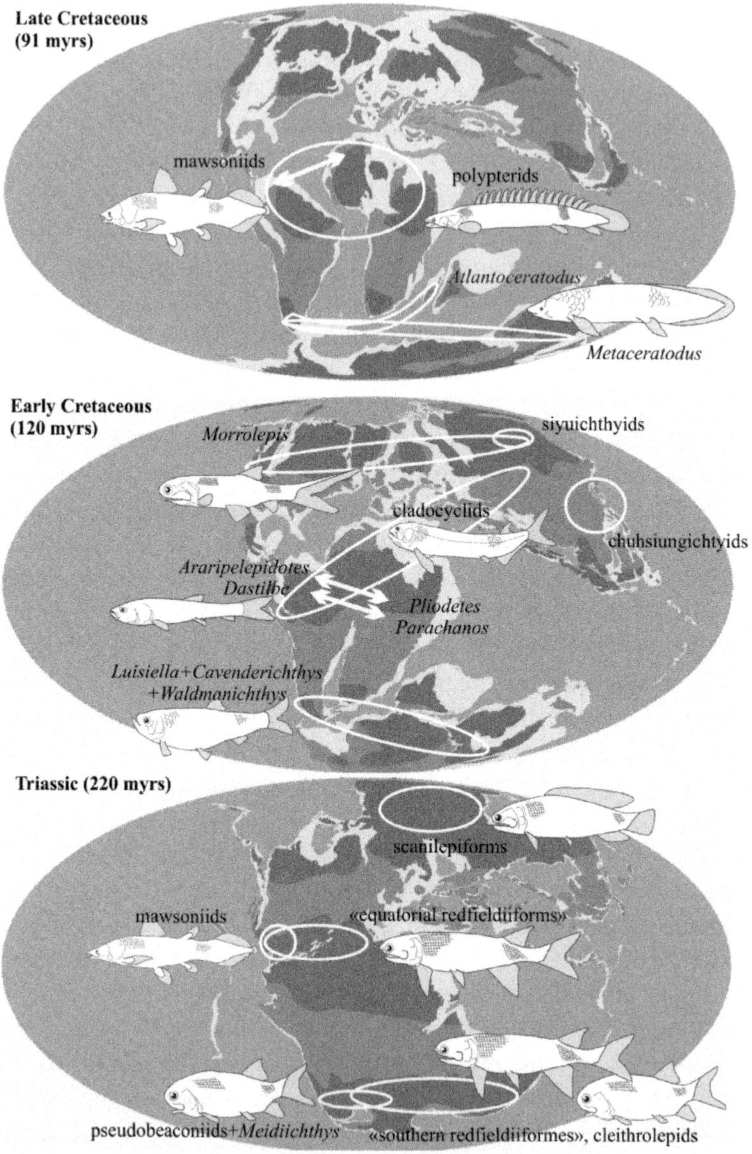

Figure 4.2. *Distribution areas of freshwater clades during the Triassic, Early Cretaceous and Late Cretaceous. The green zones correspond to humid zones, where precipitation was greater than evaporation (according to Scotese [SCO 14]). For a color version of the figure, see www.iste.co.uk/cavin/fishes.zip*

Mawsoniids were a significant part of the freshwater and brackish fauna of the Early Cretaceous and the beginning of the Late Cretaceous (Cenomanian) in western Gondwana (Africa and South America). These faunas are characterized by other groups of vertebrates. Among them, associated with the mawsoniids are basal lepisosteiforms (*Pliodetes* in Africa and *Araripelepidotes* in South America), obaichthyids present on both sides of the Atlantic and osteoglossomorphs (*Chanopsis* and *Palaeonotopterus* in Africa and *Laeliichthys* in South America). Several groups of tetrapods are consistently found in the same sites, including araripemydid turtles, araripesuchid crocodiles and, among the dinosaurs, rebbachisaurid sauropods and carcharodontosaurid and spinosaurid theropods. The latter, in particular, are very often connected with continental mawsoniids, such as forms associated with baryonycins in the Early Cretaceous (*Cristatusaurus* in Niger and *Irritator* in Brazil) and spinosaurins at the start of the Late Cretaceous (*Spinosaurus* in North Africa and *Oxalaia* in Brazil). *Spinosaurus* is now considered to be a semiaquatic theropod based on isotope analyses [AMI 10] and its skeletal anatomy [IBR 14, VUL 16]. Its almost invariable association with the remains of mawsoniids in Gondwana, and with lepisosteiforms more globally, indicates a strong connection between these dinosaurs and these fishes in specific environments, and even an ecological link between the taxa themselves.

There is little information available about the lifestyle of mawsoniids. The mouth was toothless (only small denticles line the internal part of the mandible and the palate) and opens into a large buccobranchial cavity. These features may be an adaptation to an intake of food using suction, as is the case with the present-day *Latimeria*, but it does not exclude the possibility that this morphology corresponds to a filtration feeding method. Microphagy by filtration has been suggested for giant forms of latimeriids like *Megalocoelacanthus*, the marine sister group of mawsoniids, although it has not yet been proven [SCH 09].

4.2. Lungfish (Dipnoi)

Like the coelacanths, fossils of lungfishes were known to Western naturalists before living specimens were discovered. At the beginning of the 19th century, the British doctor John Parkinson interpreted the fossilized dental plate of a lungfish as a fragment of a tortoise plastron. Louis Agassiz recognized the ichthyian and dental nature of this fossil but, due to the absence of current comparable species, he attributed it to a form closely related to a bullhead shark (*Heterodontus*). Following descriptions of living lungfishes, including the South American *Lepidosiren* in 1837, the African *Protopterus* in 1839 and the Australian *Neoceratodus* in 1876, the fossilized dental plates could be correctly interpreted. However, despite the discovery of present-day species, the affinities of these organisms were debated for several decades: were these animals amphibians, as the presence of lungs seemed to indicate, or were they fishes? Also, what was their position in relation to vertebrates with four legs, or tetrapods? While lungfishes were initially considered to be an intermediary link between fish and tetrapods, they later lost this position to the coelacanths until anatomical [ROS 81] and molecular studies at the end of the 20th Century repositioned them as the sister-group to tetrapods. The morphology of present-day lungfish, however, bears little resemblance to the first tetrapods from 365 million years ago because their evolutionary history has been very rich since this period. However, it should be noted that the low diversity of modern lungfishes and the slow morphological evolution of the post-Paleozoic forms have caused the current genera, especially *Neoceratodus*, to present the strongest "evolutionary originality" of any vertebrate [CAV 11].

Lungfishes have an original evolutionary history. Present in the fossil record since the Early Devonian, they are known throughout this entire period through marine species. In these forms, the skull is highly ossified and covered by a certain type of bony tissue called cosmine that also protects their scales. Their dentition is composed of either individualized denticles or dental plates made up of fused denticles. The taxa with dental plates diversified in the second part of the Paleozoic, during the Carboniferous and the Permian, and mainly lived in continental environments. The dental plates of lungfishes mineralized more and more over time while the rest of their skeleton, especially the cranial skeleton, showed a decrease in ossification. Over the course of this evolution, the cranial dermal bones sunk down into the dermis. One of the consequences of this evolutionary trend was a change in the nature of the fossil record of lungfishes over time: while several marine genera from the Devonian are known from articulated and relative complete skulls, the forms from the Carboniferous, Permian and especially the Meso-Cenozoic are primarily known from isolated dental plates. These plates are rarely connected to cranial elements and even more rarely to postcranial elements. One indirect consequence of this kind of preservation is that it is difficult to define characteristics that are common to all of the fossil representatives of the group and find phylogenetic relationships within the clade. Several phylogenies have been proposed for Paleozoic genera, such as in [QIA 09] and [PAR 14]. Only a few have been proposed for post-Paleozoic taxa [SCH 04, MAR 82, CAV 07a]. However, there are few overlaps between the two types of phylogenies. One question that remains unanswered is to determine whether, among the current genera, the pair *Protopterus–Lepidosiren* detached from its sister-group *Neoceratodus* during the Late Jurassic or the Cretaceous, as suggested in Schultze [SCH 04], or if the two lineages separated long before that, in the Paleozoic or at the start of the Triassic, as suggested in [CAV 07a]. In the latter case, several Mesozoic taxa would be found to have originated from the *Protopterus–Lepidosiren* clade. The choice between these two hypotheses would have significant implications on the paleoecological and paleobiogeographical interpretation of the group. Present-day lungfishes have a double branchial and pulmonary respiration system. The latter is non-essential for *Neoceratodus* but necessary for *Lepidosiren* and *Protopterus*. In addition, these two genera share the capacity to estivate buried in mud in dried up ponds (enclosed in a cocoon for *Protopterus*), while *Neoceratodus* cannot. The genus *Gnathorhiza* is known in the Permian and the Triassic of North America and Europe. Incomplete specimens have been discovered in Early Permian sites in Texas within extended sedimentary structures interpreted as estivation burrows [ROM 54]. Therefore, this ecological characteristic of present-day African and South American genera may have already appeared in the Early Permian and could represent a feature shared between them. If this is the case, then the separation between this lineage and that of *Neoceratodus* is even more ancient than previously suggested and dates back at least to the Early Permian.

The vast majority of post-Paleozoic lungfishes are found in sediment from continental areas and, for the few specimens found in marine deposits, clues suggest that they were transported there from continental environments. As mentioned, the fossils of post-Triassic lungfishes mainly consist of dental plates that are isolated or connected to the bone that supports them. The taxa are often defined based only on these plates. A dozen species, however, are known from cranial bones that are often dislocated. They are very important for understanding the relationships between Meso-Cenozoic species and, consequently, for drawing paleobiogeographical conclusions.

Lungfishes worldwide presented a relatively high taxic diversity in the Triassic that decreased until the present period [SCH 04]. In the Triassic, lungfishes were present in Laurasia and Gondwana. The genus *Ceratodus* served as a "catch-all" taxon that accommodated a large number of species with unclear affinities. As we define it here (*Ceratodus latissimus* and *C. sturii*), this genus is known in the Late Jurassic of Europe and in the Early Cretaceous in North America [SCH 81]. In the Triassic of Europe, there are also the genera *Ptychoceratodus* and *Gnathorhiza*, the latter being especially abundant during the Permian of North America. Lungfishes are more diverse in the Triassic of Gondwana: *Paraceratodus* and *Beltanodus* in Madagascar, *Arganodus* in Morocco, *Microceratodus* in Angola, and *Gosfordia* and *Ariguna* in Australia. Dental plates attributed to *Asiatoceratodus* have been recorded on almost all continents from the Triassic to the Late Cretaceous [ALV 13]. The paleogeographical status of Triassic genera is not clear [CAV 07a], with the exception of *Gosfordia* in Australia that seems to present similarities with *Paraceratodus* in Madagascar [KEM 94]. This link can be easily explained by the proximity of the two continental masses during the Triassic. During the Jurassic, lungfishes were rare in Laurasia with only *Ferganoceratodus jurassicus* during the Middle Jurassic of Kyrgyzstan [NES 85] and *Ferganoceratodus martini* during the Late Jurassic of Thailand [CAV 07a].

In the Cretaceous and the Paleogene, the biogeographical understanding of the fossil record of lungfishes remains difficult because various genera based on isolated dental plates have poorly understood relationships. In South America, *Atlantoceratodus iheringi* was found in the Coniacian levels of the Mata Amarilla Formation in southern Patagonia [CIO 07]. These authors emphasize that "*Ceratodus*" *madagascarensis* from the Campanian of Madagascar probably belongs to this genus, thus confirming the proximity already observed between the faunas of tetrapods in southern South America and Madagascar during the Late Cretaceous [KRA 06]. Agnolin [AGN 10] describes a second species of *Atlantoceratodus* in the Campano-Maastrichtian of Patagonia and Argentina. In this work, Agnolin discusses the paleobiogeographical relations of lungfishes in South America and observes that during the Cretaceous, taxa in the southern part of the continent have links to Australia and Madagascar (Figure 4.2) while taxa from the northern part of the continent have ties to Africa. Among the latter, "*Ceratodus*" *tiguidiensis* and "*Ceratodus*" *africanus*, known from the Late Jurassic–Early Cretaceous in Uruguay, the Cenomanian of Brazil, and various Cretaceous sites in Africa, are typical of western Gondwana [SOT 10]. The genus *Metaceratodus* is from the Early Cretaceous to the Pleistocene of Australia. The oldest occurrences come from the Early Cretaceous of the Griman Creek Formation in New South Wales. It is later found in the Late Cretaceous of the Winton Formation in Queensland, as well as in the Late Oligocene–Middle Miocene of Lake Palankarinna, and once again in the Pliocene of Queensland [KEM 97a]. *Metaceratodus* is also present in South America with *Metaceratodus kaopen* in the Late Santonian–Early Campanian of the Anacleto Formation in the Río Negro province, and *M. wichmanni* in the Late Campanian–Early Maastrichtian of the Río Negro, Neuquén and Mendoza provinces in Argentina [CIO 12]. The distribution of this genus, clearly freshwater, is an indication of the continental continuity between South America and Australia during the Late Cretaceous (Figure 4.2). In the Cenozoic, the fossils of lungfishes in South America belong to the genus *Lepidosiren* with the exception of one indeterminate "ceratodontid" mentioned in the Late Paleocene–Early Eocene of the Las Flores Formation in southern Argentina [CIO 10]. In Africa, we saw that the fossil record of the Early and "mid" Cretaceous shows clear links with South America. Martin [MAR 95] described the genus *Lavocatodus* in the Paleogene of North Africa and Longrich [LON 17] described *Xenoceratodus* in the Late Eocene of Libya. According to Longrich, these two genera form the family of lavocatodids, which has representatives in the Late Cretaceous in Africa and, more surprisingly, a possible representative in the Aptian of Australia. The lavocatodids evolved in Africa, with the exception of the Australian occurrence, parallel to the lepidosirenids (or the lepidosireniforms according to Longrich) present in western Gondwana. They went extinct at the end of the Eocene. The evolutionary history of *Protopterus* in Africa is detailed in Otero [OTE 11]. Among the other occurrences in western Gondwana, *Lupaceratodus useviaensis* was described from the Cretaceous Galula Formation in Tanzania [GOT 09] and *Retodus tuberculatus* in the Albian of Algeria [TAB 63, CHU 06]. Australia is the continent with the richest fossil record. The current family of neoceratodontids is represented by three genera: *Neoceratodus*, *Mioceratodus* and *Archaeoceratodus*, and by several species [KEM 05]. The genus *Archaeoceratodus* was present in a large part of Australia from the Triassic to the Miocene, *Mioceratodus* from the Eocene to the Pleistocene, and *Neoceratodus* is known as of the Early Cretaceous with *Neoceratodus nargun* in the Late Aptian and Early Albian of Cape Otway, Victoria [KEM 92, KEM 97b].

4.3. Polypterids (Cladistia)

Cladistia, or polypterids, are the most basal clade of living actinopterygians. They have primitive features like the presence of ganoid scales and a maxilla firmly fixed to the cheek. The two genera of extant polypterids, *Polypterus* or bichirs and *Erpetoichthys* or reed fish, have very derived features. The most noteworthy is a dorsal fin divided into several small fins that are each preceded by a spine. On the basis of the phylogenetic position of polypterids, the clade must be very ancient and must have detached from the rest of the actinopterygians (that is, actinopteris) in the Paleozoic, probably before the Permian. However, the fossil record for the group only begins later, at the base of the Late Cretaceous (Cenomanian). It is likely that the synapomorphies of present-day polypterids, especially their divided dorsal fins, were not present in ancient taxa. We can assume that some primitive polypteriforms have been described as "paleonisciforms", a non-natural group including fishes with primitive characters. These forms could soon be recognized as cladistia. Among the most ancient polypterids is *Serenoichthys kemkemensis* from the Cenomanian of the "Kem Kem beds" in Morocco [DUT 99]. In this species, the dorsal fin is already divided but the body is proportionally short and high, as opposed to the very long body of the bichir (*Polypterus*) and, even more so, of the reedfish (*Erpetoichthys*). Details about the skull of *Serenoichthys* are not known but the description of still-unpublished material will certainly provide more information about the osteology of this basal form, as well as the diversity and evolution of the group. In addition to this genus, which is known from articulated specimens, at least seven genera of polypterids (including the current genus *Polypterus*) were recorded in the Cenomanian of Wadi Milk in Sudan [WER 97] and in the Coniacian-Santonian of In-Becetem in Niger [GAY 88, GAY 97]. The interpretations of these genera are based on the morphology of spines of the dorsal fins (pinnules). This astonishing diversity must still be proven with complementary bony elements. The first records of polypterids in North Africa are older. Already in 1925, Stromer [STR 25] identified ganoid scales of polypterids from the Cenomanian of the Bahariya oasis in Egypt. This identification was challenged by Weiler [WEI 35] who attributed the scales to a "*Lepidotes*". Stromer then rejected [STR 36] this new identification and confirmed the attribution of these scales to a polypterid on the basis of histological cross-sections. In 1984, based on an ectopterygoid also from Bahariya, Schaal [SCH 84] described a new species of polypterid that he attributed, with some uncertainty, to the genus *Polypterus*, *Polypterus bartheli*. Later, Smith *et al.* [SMI 06] confirmed the attribution of the scales from Bahariya to a polypterid. The scales are characterized by irregular ganoid coverage forming bands between which the bony base of the scale is visible. Grandstaff *et al.* [GRA 12] referred this material to a new genus,

Bawitius, adding rare cranial elements to it. This genus was then recognized, notably because of the structure of the scales, in the Albian of the Democratic Republic of Congo [CAS 61] and in the Cenomanian of Morocco [CAV 15, MEU 16] and Algeria [BEN 15]. The remains of *Bawitius* are present in the "middle" Cretaceous in a large part of North Africa and could be abundant locally, as in Bahariya [GRA 12], Morocco (personal observations) and certain Algerian sites (M. Benyoucef and M. Adaci, personal communication, 2015). This large fish, 3 m according to Grandstaff *et al.* [GRA 12], was probably a carnivore like today's bichirs and would have been an apex predator in certain ecosystems. The genus *Polypterus* can be found in Africa since the Eocene of Libya [OTE 15] and Egypt [MUR 10a].

The ancient existence of polypteriforms, implied by their position in the phylogeny of the actinopterygians, suggests a past distribution area larger than Africa's current distribution area. Since the 1990s, Gayet and Meunier [GAY 91a, GAY 91b] have demonstrated the presence of polypteriforms in the Maastrichtian and the Paleocene of Bolivia based on characteristic ganoid scales, fragments of skulls and pinnules. Two taxa were named: *Dagetella sudamericana* and *Latinopollia suarezi*, the first being attributed to polypterids and the second to polypteriforms [GAY 92, MEU 96, MEU 98]. Based on these occurrences, it is likely that this lineage, already confined to freshwater environments, dates back to a more ancient period than the fragmentation of western Gondwana. Recently, the polypterid *Bartschichthys* was recorded in the Albian-Cenomanian of the Alcântara Formation in northeastern Brazil [CAN 11]. This genus was previously mentioned from the Late Cretaceous in Niger and from the Cenomanian of Morocco. This means that the distribution of polypterids covered a significant part of western Gondwana during the Late Cretaceous (Figure 4.2). It is likely, however, that the representatives of the cladistia could be discovered in older deposits than the Cretaceous. From the beginning of the fossil record of this clade, polypterids were present in sediment from freshwater environments. However, it is reasonable to consider that the group originated in a marine environment as indicated by the reconstruction of ancestral environments [BET 15] (or, for a different point of view, see [VEG 12]). The transition from marine to freshwater environment probably occurred at the end of the Paleozoic or during the Triassic period.

4.4. Coccolepidids

This extinct family of chondrosteans was present in both marine and freshwater environments. Skrzycka [SKR 14] discussed the evolution of this family. The most ancient representatives are the Early Jurassic marine species *Coccolepis liassica* from Lyme Regis in England and the freshwater species *Plesiococcolepis hunanensis* from Hengnan in China. *Coccolepis* was still present in marine environments during the Late Jurassic with *Coccolepis bucklandi* in Solnhofen, Germany [HIL 04], as well as in freshwater with *Coccolepis australis* during the Late Jurassic and *Coccolepis woodwardi* during the Early Cretaceous in Australia. In Laurasia, the genus *Coccolepis* was present during the Early Cretaceous with *Coccolepis yumenensis* in China and *Coccolepis macroptera* in Bernissart, Belgium. The genus *Condorlepis* was present in the continental Late Jurassic of Argentina [LÓP 13]. *Yalepis*, from the Middle Jurassic of Asia, is attributed with some uncertainty to the coccolepids. Our lack of understanding of the relationships between the different species of *Coccolepis* and between the genera within the family, as well as their apparently considerable tolerance for variations in salinity, does not allow us to pinpoint paleobiogeographical signals. *Morrolepis* is another genus with an interesting distribution. It was found in the Late Jurassic of the Karabastau Formation in Kazakhstan, in the Late Jurassic of the Morrison Formation in the United States, and, according to Skrzycka [SKR 14], in the Purbeckian of England (the latter species was previously attributed to the genus *Coccolepis*) (Figure 4.2). *Morrolepis* had a distribution that extended across the entire northern hemisphere, much like the current distribution of the sturgeon *Acipenser*. Skrzycka [SKR 14] noted that the primitive actinopterygians, among them the coccolepidids, dominated the freshwater ichthyofaunas in the Late Jurassic of Karatau where derived actinopterygians (for the Jurassic) were rare. This feature is found in several freshwater assemblages in the Mesozoic and Cenozoic.

4.5. Sturgeons and related fishes (Acipenseriformes)

Today, acipenseriforms include sturgeons (acipenserids), distributed throughout the northern hemisphere, and paddlefish (polyodontids), limited to a few river basins in the United States and southern China. The skeletons of these fishes are mainly cartilaginous, hence the name chondrosteans. However, the cartilaginous nature of these fishes is a reversion from a bony skeleton and does not indicate a link with true "cartilaginous fishes" like rays and sharks (chondrichtyans). Like the polypterids, the origin of this group must be ancient and certainly dates back to the Paleozoic, but the fossil record does not begin until the Early Jurassic with the marine genus *Chondrosteus*. Prior to that, in the Middle Triassic, a freshwater chondrostean belonging to an indeterminate order, *Neochallaia*, was found in the Cuyana basin in Argentina [LÓP 10]. From the Early Cretaceous, all acipenseriform fossils are of continental origin. Therefore, it is probable or at least reasonable to interpret the diadromous lifestyle of the present-day genera *Acipenser* and *Huso* as a partial return to the marine environment from a strictly freshwater lifestyle. The diversity of the order is relatively high in the Jiehol biota during the Early Cretaceous in Liaoning and surrounding regions, where there are several genera of a family endemic to East Asia, the peipiaosteids (*Liaosteus, Peipiaosteus, Yanosteus, Boroichthys, Stichopterus*) (Figure 4.3).

The most ancient polyodontid known is *Protopsephurus* present in the Early Cretaceous of the *Lycoptera* fauna in China [GRA 02]. Chang and Miao [CHA 04] recorded the presence of this genus in southern Siberia accompanied by another potential polyodontid, *Alexandrichthys*. Among the other extinct genera of polyodontids are *Paleopsephurus*, from the Maastrichtian of the Hell Creek Formation in Montana, USA, and *Crossopholis*, from the Eocene of the Green River Formation in Wyoming, USA. The genus *Polyodon* appeared early with the species *Polyodon tuberculata* in the Early Paleocene of the Tullock Formation in Montana [GRA 91]. The oldest mention of an acipenserid comes from North America. It is a recently described fragment from the Cenomanian of the Dunvegan Formation in northwestern Canada [VAV 14]. This skull fragment comes from a fish that was about 5 m in total length. It is in this group that we find the largest freshwater fishes today, with beluga (*Huso huso*) individuals capable of reaching eight meters in length. In the terminal Cretaceous, the isolated remains of acipenserids are relatively abundant in North America, but only two genera are sufficiently well known to have been named. These are *Psammorhynchus longipinnis* from the Judith River Formation in Alberta, Canada, and *Protoscaphirhynchus squamosus* from the Hell Creek Formation in Montana, USA [GRA 06]. The authors note that a few well-preserved specimens from these sedimentary deposits have been found in the abdominal cavities of the carcasses of hadrosaurian dinosaurs. They suggest that this "microenvironment" was favorable for the conservation of fragile skeletons of these fishes in an environment subject to strong energy. Grande and Hilton [GRA 06] list all fossils of acipenserids except occurrences in the

Pleistocene. A few isolated elements attributed to this family are known in the Late Cretaceous in Uzbekistan and Belgium, in the Paleocene of Kazakhstan and Montana, in the Eocene of England and France and in the Oligocene of France. In the Neogene, the family was found in Europe, the United States and Japan.

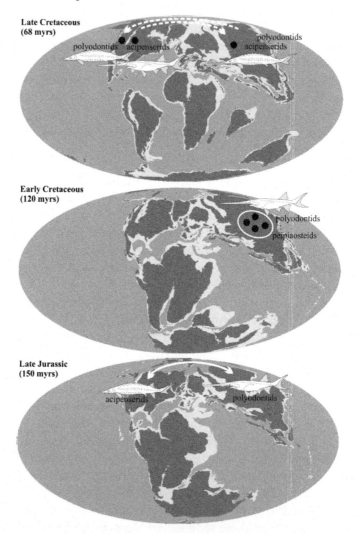

Figure 4.3. *Evolutionary history of acipenseriforms during the Mesozoic. The vicariances are represented by double-ended arrows, dispersals by dotted arrows and occurrences by stars. For a color version of the figure, see www.iste.co.uk/cavin/fishes.zip*

According to the phylogeny of polyodontids by Grande *et al.* [GRA 02], the Asian genus *Protopsephurus* is the most basal, which indicates that the family originated on that continent (Figure 4.3). The most ancient acipenserid being North American, we can suggest a vicariance between the two families on either side of Beringia, an event that was followed by dispersals in both directions: polyodontids moved toward North America and acipenserids moved toward Asia. This scenario is based on only a small number of fossils and could easily be invalidated by new discoveries. However, it is clear that Beringia was a continental passage during a large part of the Cretaceous for acipenseriforms as for other groups of freshwater fishes (hiodontids, catostomids, ellimmichthyiforms). More detailed discussions about the biogeographical history within the acipenserids were proposed by Peng *et al.* [PEN 07] and Laumann [LAU 16].

4.6. "Paleopterygians" and "sub-holosteans"

These two taxa are not monophyletic and include a wide variety of genera from the end of the Paleozoic and from the Triassic. However, these groups are important because they are overrepresented in freshwater assemblages in the Triassic compared to neopterygians [ROM 16]. As examples of genera, many of which should be revised, we note around 10 in Australia and South Africa including *Elonichthys, Leighiscus, Pristisomus, Urosthenes, Zeuchthiscus, Tripelta, Dicellopygae* and *Ischnolepis*, as well as *Schaefferius* and *Dimorpholepis* in Angola; *Endemichthys* in Lesotho; *Eoperleidus, Oshia* and *Evenkia* in Siberia; *Ferganiscus* and *Sixtelia* in Kyrgyzstan; *Duwaichthys* and *Sinkiangichthys* in Xinjiang; *Mizhilepis, Triassodus* and *Wayaobulepis* in Shaanxi; *Shuniscus* in Sichuan; *Hyllingea* in Sweden; *Luederia, Rushlandia, Tanaocrossus* and *Turseodus* in the United States; and *Woodthorpea* in England [SCH 67, FOR 73, LOM 01, LÓP 04, JIN 06, MOG 09, ROM 16]. Some of these genera have majority of marine species with just a few freshwater representatives. This is the case with *Australosomus*, which may be a basal neopterygian according to Xu and Gao [XU 11] and Xu *et al.* [XU 14]. It was found in the freshwater deposits of the Early Triassic of Kenya and Tanzania. *Boreosomus* was found in the Early Triassic of Kenya, the Middle Triassic of the United States and the Late Triassic of Shaaxi, China [JIN 06, ROM 16]. *Gyrolepis* was found in the Late Triassic of China, *Palaeoniscum* in the Middle Triassic of Australia and the Late Triassic of China. *Ptycholepis* was found in the Late Triassic of the United States and Madagascar. These genera can be considered to be euryhaline, although a systematic revision could reveal deeper differences between the marine and freshwater forms. Beyond these paleoecological observations, it is not possible to trace the evolutionary history of these groups or ascertain paleobiogeographical signals without phylogenetic analyses. The information that we learn from "paleopterygian" and "sub-holostean" fishes is that primitive forms were diversified and endemic in freshwater during the Triassic in many regions of the world, especially Australia, South Africa, central Asia and North America.

4.7. Redfieldiiforms (Redfieldiiformes)

Redfieldiiforms are a clade of freshwater fishes from the Triassic and Early Jurassic with unclear affinities within the basal actinopterygians. They have been found in the Middle Triassic of southern Gondwana in South Africa (*Atopocephala, Daedalichthys, Helichthys*) and Australia (*Brookvalia, Phlyctaenichthys, Geitonichthys, Molybdichthys* and *Scizurichthys*), in the Middle Triassic of China (*Sinkiangichthys*), Middle Triassic of Germany, and Late Triassic of Silesia in Poland (*"Dictyopyge"*), as well as in the Late Triassic and Early Jurassic of the Newark Formation in the United States (*Dictyopyge, Cionichthys* and *Redfieldius*) and finally in the Late Triassic of the southwestern United States (*Cionichthys, Synorichthys, Lasalichthys*) [SCH 84, LÓP 04, JIN 06, ROM 16]. Redfieldiiforms diversified in freshwater at the beginning of the Mesozoic and their distribution should allow us to pinpoint paleogeographical signals to the extent that their relationships to each other can be partly resolved. Martin [MAR 82] demonstrated a direct biogeographical relation between North Africa and North America in the Late Triassic by showing the proximity between *Mauritanichthys*, from the Fergana corridor in Morocco, and *Lasalichthys* from North America. Later, in 1984, Schaeffer [SCH 84] revised the Redfieldiiforms and proposed a new phylogeny in which the basal genera, in pectinate position relative to each other, were all South African or Australian, whereas the more derived genera were North American or North African. Although the uncertain status of certain genera encourages us to interpret this information with care [LOM 13], it also indicates two zones of endemism within Pangaea in the Triassic: one located in southern Gondwana, which includes "southern redfieldiiforms", and another located in North America and North Africa, which includes "equatorial redfieldiiforms" (Figure 4.2).

4.8. Scanilepiforms (Scanilepiformes)

This stem or basal order of neopterygians was recently discussed by Xu *et al.* [XU 14] in the framework of the revision of *Fukangichthys*. The genera included into this order are *Fukangichthys, Beishanichthys, Evenkia* and *Scanilepis*, which are all freshwater fishes except the last one. *Evenkia* was discovered in the Early Triassic of the Tunguska basin in Siberia, *Beishanichthys* in the Early Triassic of Gansu in China and *Fukangichthys* in the Middle Triassic of Xinjiang, also in China. This concentration of closely related taxa seems to correspond well to a small radiation in central Asia (Figure 4.2). Schoch and Seegis [SCH 16] mentioned the presence of scanilepidids in the Middle Triassic of southern Germany, which increases the area of the clade's distribution toward the west.

4.9. Perleidiforms (Perleidiformes)

Although the list of taxa that make up this group and their phylogenetic relationships are not yet clearly defined, the first cladistic analyses by López-Arbarello and Zavattieri [LÓP 08] and Sun *et al.* [SUN 12] have interesting paleobiogeographical conclusions. In this work, we will only address the continental taxa and must therefore exclude the marine family of perleids. The cleithrolepid family includes *Hydropessum* and *Cleithrolepidina* in the Middle Triassic of South Africa, with *Cleithrolepis* in the Early and Middle Triassic of Australia confirming the strong connection between these two regions (Figure 4.2). López-Arbarello and Zavattieri [LÓP 08] defined the pseudobeaconiid family for two South American genera, *Pseudobeaconia* and *Mendocinichthys*, whose sister genus is *Meidiichthys* in South Africa, underscoring the link between these two parts of western Gondwana (Figure 4.2). *Dipteronotus*, present in the continental Middle Triassic of Germany and the continental Late Triassic of Morocco constitutes, with *Felberia*, the sister group of the cleithrolepids. Insofar as *Felberia* and more ancient occurrences of *Dipteronotus* are marine, it is reasonable to consider this pair of genera as a return to a marine environment for this otherwise freshwater clade in connection with a dispersal toward Europe. Freshwater perleidiforms are present in the Early Triassic of the Cassanga series in Angola with *"Perleidus" lutoensis* and *"Perleidus" lehmani*. Inclusion of these African forms in a phylogeny of the order will make it possible to identify the biogeographic links for this fauna, which is otherwise characterized by several endemic elements [ANT 90].

4.10. Holosteans (Holostei)

Holosteans are a taxon created by Müller in 1846 [MÜL 46] that today includes three genera of freshwater fishes from North America that are described as primitive: gars (*Lepisosteus* and *Atractosteus*) and bowfin (*Amia*). The taxonomic history of this group is complex. Many fossil forms, including the semionotiforms, were associated with gars, also called ginglymodians, in the 19th and 20th Centuries. Then, the majority of these extinct taxa were removed from the clade following new research, especially by Patterson [PAT 73]. Many other fossil forms from the Mesozoic were also referred to the second group, called halecomorphs. The direct relation between ginglymodians and halecomorphs (that is holosteans) was questioned by Patterson's research. According to him, and the majority of ichthyologists and paleoichthyologists for about 40 years after, halecomorphs were considered to be closer to teleosteans than to ginglymodians. Halecomorphs and teleosteans make up the "halecostom" clade. Living halecostoms actually share a certain number of synapomorphies that are absent from ginglymodians, such as the presence of an interoperculum and a mobile maxilla. However, a recent description of ginglymodians belonging to the obaichthyid family from the "mid" Cretaceous in Brazil revealed the presence of features supposedly absent from this group, including an interoperculum and a mobile maxilla [GRA 10]. The analysis of this new distribution of characters in extant and fossil forms made it possible to identify the monophyly of holosteans, as Grande's study initially showed [GRA 10]. It should be noted in passing that the "resurrection" of holosteans based on morphology relies on the study of fossils showing unique combinations of characters. This case study underscores the role that extinct forms, known only from fossil forms, can play in our understanding of relationships between present-day clades. This view was previously challenged by Patterson [PAT 81] who regarded the message of fossils as useless in solving phylogenetic relationships. Since the turn of the 20th Century, several molecular phylogenetic analyses have identified the monophyly of holosteans. Ginglymodians and halecomorphs are therefore treated in the same chapter but in separate sections because they correspond to distinct episodes of incursions of initially marine clades into freshwater environments.

4.10.1. *Ginglymodians (Ginglymodi)*

The two genera and seven species of present-day ginglymodians in North and Central America, the gars, make up a well-defined group of fishes that share several unique derived characters. Unique among bony fishes with rayed fins, or actinopterygians, ginglymodians do not have vertebrae with two concave articular faces (amphicoelous), but rather vertebrae with a convex anterior face and a concave posterior face (opisthocoelous). Another particularity of these carnivorous fish is their long snout with teeth made of plicidentine, a very strong infolded dentine. One feature of these fishes that is primitive, not derived like the above-mentioned features, is the presence of thick ganoid scales. These scales are made of a bony basal plate covered by a layer of enamel. The fossil record of the most derived ginglymodians, the lepisosteids, was reviewed in a very detailed way by Grande [GRA 10]. If we go back through the fossil record of this family, we observe that occurrences of lepisosteids are abundant in the Neogene of North America where the clade has been confined since this period [SMI 81]. In the Paleogene, the fossil record is relatively abundant in North America and Europe. The last European representatives date from the Oligocene with "*Lepidosteus bohemicus*" in the Middle Oligocene of the Czech Republic and "*Lepidotus fimbriatus*" in the Eocene and Oligocene of England, Belgium and France, known from a large number of isolated elements. These remains are attributed to indeterminate lepisosteids by Grande [GRA 10]. The lepisosteids even survived in Europe until the Miocene, as attested by an isolated scale attributed to "*Lepidosteus strausi*" from the Early Miocene of the Frankfurt region of Germany [GRA 10]. When the fossil record in North America and Europe permits a more precise identification, the fossils are referred in large part to the lepisosteins, that is a clade that includes the two present-day genera *Atractosteus* and *Lepisosteus* as well as the extinct genus *Cuneatus*. In Eocene deposits from Messel, Germany, and Green River in the United States, one particular gar, *Masillosteus*, is included in its own subfamily beside the lepisosteins. It does not have the long snout or strong conical teeth typical of modern piscivorous gars, but an array of small rounded teeth probably adapted for a molluscivorous diet [MIC 01, GRA 10]. This genus lived in sympatry with other gars that had morphologies and diets that are more common for this family [GRA 10].

In the Late Cretaceous, lepisosteids are known primarily from isolated fossils such as teeth, scales and skeletal fragments that are abundant in some North American and European deposits. Most are referred to taxa that are considered non-valid by Grande, such as *"Lepidotus haydeni"* and *"Lepidotus occidentalis"* from the Campanian of the Judith River Formation in Montana, *"Lepisosteus opertus"* from the Maastrichtian of the Hell Creek Formation in Montana and *"Clastes lusitanicus"* from the Late Cretaceous in Portugal. The difficulty in identifying these isolated remains, especially ganoid scales, led Mireille Gayet and François Meunier to search for diagnostic features based on the micro-ornamentation of the enamel that covers the scales. This is how they demonstrated that each species has a unique distribution pattern of microtubercles on the surface of the scales characterized by their size and density [GAY 86, GAY 88, GAY 02, MEU 96]. The lepisosteid family is present in the Late Cretaceous outside of North America and Europe. *Lepisosteus indicus*, for example, is present in the terminal Cretaceous of the Lameta Formation in Madhya Pradesh, India [GRA 10]. In Madagascar, the remains of indeterminate lepisosteids were found in the probable Campanian of the Maevarano Formation in the Mahajanga basin [GOT 98]. In Africa, *"Paralepidosteus praecursor"* was described in the basal Cretaceous (Neocomian) of central Africa [CAS 61] and *"Paralepidosteus africanus"* was recorded in the Late Cretaceous (Santonian) of Niger [ARA 43]. The latter species was also found in the Campanian of the south of France [CAV 96]. Grande [GRA 10] considered *"P. africanus"* to be a *nomen dubium* and attributed the holotypic material to an indeterminate lepisosteid, like the material of *"P. praecursor"*. Gayet and Meunier [GAY 01], however, consider the genus *Paralepidosteus* to be valid on the basis of the micro-ornamentation of the scales, but Grande did not discuss this feature when he rejected the validity of the genus. *Oniichthys falipoui*, a species from the Cenomanian of the Kem Kem beds in Morocco [CAV 01], was attributed to the genus *Atractosteus* by Grande [GRA 10] before reintegrating its original genus because of the description of new material [CAV 15]. Lepisosteids were also present in the Late Cretaceous of South America, with *"Lepisosteus"* *cominatoi* in the Baurú Formation in Brazil considered to be an indeterminate representative of the family by Grande [GRA 10]. A similar situation occurred with *"Atractosteus turanensis"* from the Turonian–Coniacian of Uzbekistan. These isolated elements do not allow us to establish phylogenetic links between these different Cretaceous taxa and, consequently, to retrace a paleobiogeographical scenario. However, they indicate that the representatives of the lepisosteid family were present on almost all continents during the Cretaceous. The oldest occurrences in the Early Cretaceous are found on the African continent and it should be noted that it is also on this continent, as well as South America, that the closest relatives of the "obaichthyids plus lepisosteids" clade are found (see below).

The origin of the lepisosteids and the phylogenetic relationships of this family with extinct taxa remained unclear for a long time. The discovery of the obaichthyids cleared it up in part. The first specimens, attributed to two distinct species of the genus *Obaichthys*, were discovered in the Albian deposits of the Santana Formation in Brazil and were described by Wenz and Brito [WEN 92]. Then, an exhaustive study of the material made it possible to incorporate one of the species into a second genus, *Dentilepisosteus* [GRA 10]. The two genera were also found in the fauna of the Kem Kem beds in Morocco [GRA 10, CAV 15]. Characterized by an ornamentation made up of strong ridges and spines of enamel, the isolated scales of obaichthyids are easily recognizable. That is how isolated scales attributed for nearly a century to *Stromerichthys*, a genus with uncertain affinities, can now be attributed to the obaichthyids [CAV 15]. This reinterpretation made it possible to increase the geographic distribution of the family to central Africa, to Egypt, to France and Portugal. Obaichthyids are ginglymodians that are closely related to the lepisosteid family, as attested by their opisthocoelous vertebrae. However, contrary to the lepisosteids, obaichthyids still have an interoperculum, an articulated maxilla and do not yet present the peculiar structure of the cheek and upper jaw of modern gars. Their mouths, which are very particular, are made up of a mandible that is considerably shorter than their moderately long rostrum. The mandible only contains teeth on the anterior end, which face a lateral process of the rostrum. This arrangement certainly does not correspond to the predatory diet of modern gars. The deposit environments where obaichthyids have been found indicate that they must have been able to tolerate the variable salinity of brackish, closed marine, deltaic or mangrove-type environments.

During the Mesozoic, several actinopterygians had ganoid scales that are superficially similar to those of gars. These "ganoides", as Louis Agassiz called them, form a heterogeneous group whose relationships are not well understood. Within this group, two genera, *Semionotus* and *Lepidotes*, have long been "catch-all" taxa that contain several poorly defined species. Recent works have made it possible to better characterize part of the species in these two genera and indicate that some of them are located at the base of the lepisosteid lineage and grouped with the latter within the lepisosteiforms. This includes two close genera, *Araripelepidotes* and *Pliodetes*, from the Albian of Brazil and the Aptian of Niger, as well as genera from the Late

Jurassic or basal Cretaceous in southeastern Asia (*Thaiichthys*, *Isanichthys*, *Khoratichthys*), all of freshwater or brackish origin, and eventually the more basal marine genera *Scheenstia* and *Lepidotes*. This series of basal lepisosteiforms illustrates the transition from marine to continental environments with intermediary forms in brackish environments.

The result of this overview, based on observations that are partly still uncertain and subject to modifications, is the following schema. Lepisosteiforms evolved from marine forms during the Jurassic. They went through a small diversification during the Late Jurassic and the Early Cretaceous in fresh water in what is now southeastern Asia. Then, in western Gondwana, we observe vicariance events between the genera *Pliodetes* and *Araripelepidotes* and between species within the obaichthyid genera when the South Atlantic opened (Figure 4.2). At the same time, in the Early Cretaceous, lepisosteids diversified on all continental masses. In the Paleogene, they were only present in Europe and North America, and finally, in the Neogene, only in North America.

Before the lepisosteiforms appeared, ginglymodians already had representatives in continental environments in the Early Cretaceous, including "*Lepidotes*" in Las Hoyas and *Neosemionotus* in South America (two genera whose relationships to the lepisosteiforms are still uncertain). Before this, *Semionotus* and *Lophionotus* had already diversified in North America in the Late Triassic and Early Jurassic. In future research, it would be interesting to observe if the transfers from marine to continental environments were rare or common events over the evolutionary history of the clade during the Mesozoic. Ginglymodians may have held the same place during the Mesozoic that the otophysans occupied in the Cenozoic in freshwater environments, but for the moment, this history is still largely unknown.

4.10.2. *Halecomorphs (Halecomorphi)*

Today, halecomorphs are only represented by the bowfin, *Amia calva*, which lives in freshwater in the eastern United States. Halecomorphs were very diversified during the Mesozoic and especially present in marine environments during the first half of that era. Some early incursions into freshwater were observed, such as *Caturus* in the continental Triassic of Argentina [BOG 13]. An excellent monograph about the amiids was published by Grande and Bemis [GRA 98]. Most of the information presented here came from this study, with a few updates. According to current data, it seems that a single transfer from marine to continental environments occurred with the amioids (Amioidea). This transition occurred in the Early Cretaceous based on the fossil record but, if we take into account the existence of ghost lineages that are implied by the phylogeny, the passage probably occurred at the end of the Jurassic. According to this schema, some amioids then returned to a marine environment. The reality of these returns, rather rare in the evolutionary history of the actinopterygians, is not very well supported because it rests on phylogenies that will still undergo important changes. However, the return to a marine environment seems well documented for two genera, *Solnhofenamia* and *Pachyamia*, but it is more problematic for *Tomognathus* due to the uncertainty surrounding its phylogenetic position [FOR 06, CAV 07b, CAV 12]. Another genus, *Calamopleurus*, was probably euryhaline as indicated by its presence in a deposit from a closed marine environment in the Santana Formation in Brazil. However, the occurrence of *Calamopleurus* in the Brazilian Ilhas and Codo formations [MAI 00] shows that it could survive in completely freshwater environments. One species attributed to the genus *Calamopleurus*, *C. africanus*, was described by Forey and Grande [FOR 98] from the Kem Kem beds of Morocco, a location that probably corresponds to a deltaic environment. Because these localities were relatively close in time, probably the Albian for the South American deposits and the Cenomanian for the African deposits, this pattern represents a good example of vicariance connected to the opening of the South Atlantic.

Within the amioids, the first intracontinental vicariance occurred between the sinamiids in East Asia, a family that is very well distinguished morphologically, and the amiids in the rest of the world (Figure 4.4). The vicariant event may have occurred more precisely between Asia and Europe because the most basal amiid, *Amiopsis*, is present in Early Cretaceous sites in Belgium (Bernissart) and Spain (El Montsech and Las Hoyas). The sinamiids were recognized for the first time in China by Erik Stensiö with the species *Sinamia zdanski*. One distinct feature of the family is the presence of an unpaired parietal. Several other species and the genus *Ikechaoamia* were later found in the *Lycoptera* fauna in northern China and in the *Mesoclupea* fauna in southern China (cf. [CHA 04] for an overview), as well as in South Korea [YAB 06], Japan [YAB 14] and Thailand with the genus *Siamamia* [CAV 07b]. Around 10 species are known in the family and the first phylogenetic analyses by Peng *et al.* [PEN 15] made it possible to outline the relationships between the biogeographical provinces in East Asia during the Early Cretaceous.

The biogeographical pattern within the other amioid family, the amiids, is not very clear, at least for the first part of their evolutionary history. One of the subfamilies, the amiins, were limited to Laurasia at that time, but the other subfamily, the vidalamiins, occupied a larger distribution area. According to Grande and Bemis [GRA 98], and confirmed by Brito *et al.* [BRI 08], the distribution of vidalamiins can be explained like this (Figure 4.4): on the one hand, the calamopleurini tribe, limited to the southern hemisphere and showing a vicariance on both sides of the South Atlantic, and on the other hand, the vidalamini tribe, presenting a coastal distribution all along North America, South America and Europe. According to this schema, certain genera were then "captured" in freshwater environments on the different continents; *Cratoamia* in South America and *Vidalamia* in Europe in the Early Cretaceous, then *Melvius* in North America during the Late Cretaceous. *Maliamia*, from the Eocene of Mali, could represent a case of "capture" of a calamopleurini in freshwater in Africa at an indeterminate moment in the Late Cretaceous or at the start of the Paleogene. This schema should also include the presence of an indeterminate vidalamiin at a high latitude in the Canadian Arctic in the Late Cretaceous [FRI 03].

The most ancient amiins are known from indeterminate genera and species in Uzbekistan, on one hand from the Late Cenomanian of the Khodzhakul Formation (*"Amia semimarina"*), where the deposit was probably freshwater, and on the other hand from the Turonian–Coniacian of the Taikarshin Member (*"Amia limosa"*), where the deposit environment has not been ascertained (Figure 241 in [GRA 98]). The only other occurrence of an amiin in the Cretaceous comes from the Turonian–Coniacian of the Canadian Arctic [FRI 03]. In the Paleogene, the amiins were represented by the genera *Amia* and *Cyclurus*. *Cyclurus* was present in the entire northern hemisphere, with occurrences in Asia, Europe and North America. *Amia* was present in North America and Europe, but occurrences of amiins *incertae sedis* in Asia lead us to believe that this genus was also present there [CHA 10]. The distribution of amiin fossils attests to dispersals of *Cyclurus* and *Amia* between the different continents of the northern hemisphere. The dispersals possibly even occurred across the Arctic Ocean, which had a low salinity during part of the Eocene [WAD 08]. The amiins disappeared from

Europe during the Middle Eocene and from Asia during the Late Miocene (Figure 241 in [GRA 98]). Occurrences of amiids in the Neogene of North America were compiled by Smith [SMI 81].

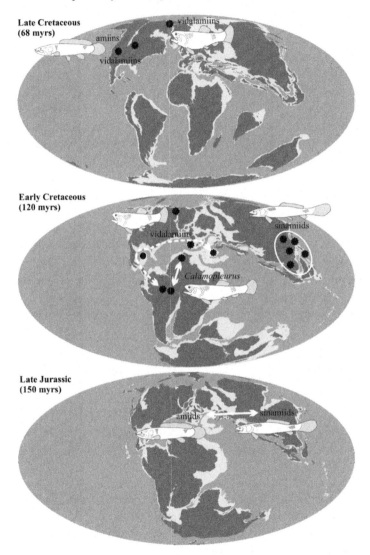

Figure 4.4. *Evolutionary history of the amioids during the Mesozoic. Vicariances are represented by double-ended arrows, dispersals by dotted arrows and occurrences by stars (stars with a white border are marine occurrences). For a color version of the figure, see www.iste.co.uk/cavin/fishes.zip*

4.11. Basal teleosteomorphs (Teleosteomorpha) and *incertae sedis*

The archaeomenids make up a poorly defined family [TAV 11] that was mainly Gondwanan with Jurassic genera discovered in the Antarctic (*Oreochima*), Australia (*Archaeomaene* and *Madariscus*, which are generally associated with *Aphnelepis* and *Aetheolepis*) and the genus *Wadeichthys* from the Early Cretaceous in Australia. The only Laurasian genus is *Zaxilepis* from the Early Jurassic of China. Insofar as the monophyly of the family has not been proven, it is too soon to use this as evidence of a link between Gondwana and eastern Asia. At most, this family represents a new potential example of diversification of a continental fish group in Gondwana.

A group of three species from the Late Jurassic of Patagonia in Argentina (*Luisiella feruglioi*), Tralbagar in Australia (*Cavenderichthys talbragarensis*) and from the Early Cretaceous of Koonwarra in Australia (*Waldmanichthys koonwarri*) are very useful for demonstrating that there were still direct links between South America and Australia in the Late Jurassic [SFE 15] (Figure 4.2). Taverne [TAV 01] connected *Paraclupavus caheni* from the Middle Jurassic of Stanleyville to *Cavenderichthys*. This possible connection represents a more northern occurrence of this clade relative to other members.

The siyuichthyids are "pholidophoriforms" *sensu largo* described by Su Dezao [SU 85] from Xinjiang, western China (Figure 4.2). The family, which is composed of poorly preserved specimens and should be revised [ARR 13] may be related to the catervariolids [TAV 11a]. These seven species, if they turn out to be valid, could represent a small evolutionary radiation of basal teleosteans in Asia equivalent to the archaeomenids in Gondwana.

The catervariolids, included in their own order of catervarioliforms by Taverne [TAV 11b], includes three genera from the Middle Jurassic of central Africa. Like the two previous families, the clade seems to represent a small evolutionary radiation of basal teleosteans in Africa.

The ankylophorids, also included in their own order by Taverne [TAV 13], are known by a large number of genera from the Jurassic, but the majority of them are marine. The two freshwater genera from the Middle

Jurassic of the Democratic Republic of Congo do not seem to form a clade, according to Taverne [TAV 13].

The same applies to the pleuropholids, the majority of whose representatives are marine but which contains a few freshwater species. These include, in central Africa, one species of *Pleuropholis* and some genera, such as *Parapleuropholis* and *Austropleuropholis* from the Middle Jurassic; in South America, *Gondwanapleuropholis* from the Late Jurassic of Brazil and some indeterminate pleuropholids from the Early Cretaceous of Argentina; and in Europe, one species of *Pleuropholis* from the Early Cretaceous of Spain. The relationships between these genera need to be better understood before drawing biogeographical conclusions.

Aspidorhynchiforms, relatively common in the Late Jurassic and Early Cretaceous, are primarily marine until the base of the Late Cretaceous (Cenomanian). The genus *Vinctifer* seems to be present in freshwater, or at least brackish, environments in the Early Cretaceous in South America. Then, in the Late Cretaceous, the last representatives of this order included in the genus *Belonostomus* were exclusively present in freshwater in North America (the recent mention of *Belonostomus* by Blanco *et al.* [BLA 16] in the Maastrichtian of Spain must be confirmed). Their presence in continental environments is an example of the refugia role of these environments for certain groups of fishes. Aspidorhynchiforms disappeared at the end of the Cretaceous, probably during the extinction at the Cretaceous–Paleogene boundary. They are one of the rare examples of freshwater fishes, together with the mawsoniid coelacanths, that died out during this mass extinction (cf. 5.3.2).

The pycnodontiforms form a well-identified group present in the fossil record from the Triassic to the Eocene. The position of this clade within the actinopterygians is uncertain. Recently, Poyato-Ariza [POY 15] addressed this question and found the pycnodontids to be a sister group of a clade made up of teleosteomorpha and holosteans. According to him, they are the most basal neopterygians. These fishes were primary marine, although there were a few exceptions. Among them, the best documented comes from Spain, in the Late Berrasian–Early Valanginian of El Montsec with *Ocleodus*, and from the Late Barremian of Las Hoyas with *Stenamara* and *Turbomesodon*. According to the phylogenies of Poyato-Ariza and Wenz [POY 02], Kriwet [KRI 05] and Ebert [EBE 16], these taxa do not make up a clade. Only the research of Poyato-Ariza and Wenz [POY 04] finds *Stenamara* and

Turbomesodon to be sister genera. This last genus includes *Turbomesodon praeclarus* in Las Hoyas in a freshwater environment, *T. bernissartensis* in Bernissart in a freshwater or brackish environment, and *T. relegans*, from the Late Jurassic of Solnhofen in a strictly marine environment. Because *T. relegans* and *T. praeclarus* form the sister clade to *T. bernissartensis*, we should consider that the *Stenamara–Turbomesodon* clade, if it can be confirmed, was originally freshwater and was followed by a return to a marine environment by the species from Solnhofen (although it is the most ancient of the three). The phylogeny of Ebert [EBE 16], however, contradicts this scenario by placing a whole series of marine taxa between *Turbomesodon* and *Stenamara*. To my knowledge, the most ancient presence of pycnodontids in a continental environment dates back to the Late Jurassic with one discrete occurrence in the Morrison Formation in the United States [KIR 98]. The occurrences, generally in the form of isolated teeth, are more abundant in sites of continental origin in the Cretaceous. These sites often correspond to environments close to marine areas. Pycnodontids were also found in the Yacoraite Formation from the terminal Cretaceous of Argentina, in the Santonian of the Iharkút site in Hungary [KOC 09] and in the Eocene of *Tamaguélelt* in Mali. In this last site, which may not be strictly freshwater, a small radiation of the genus *Pycnodus* with three species was observed [LON 84].

4.12. Ichthyodectiforms (Ichthyodectiformes)

The ichthyodectiforms form an order of basal teleosteans that disappeared at the Cretaceous–Paleogene boundary. Although primarily marine, these fishes do have a few representatives in brackish and freshwater environments.

The Chuhsiungichtyids are an Early Cretaceous family defined by Yabumoto in 1994 [YAB 94] that includes the genus *Chuhsiungichthys* initially described in Yunnan and then in Japan, the genus *Mesoclupea* initially described in Zhejiang and then in various sites in southern China, and finally *Jinjuichthys* from the Jinju Formation in South Korea [KIM 14] (Figure 4.2). This family is typical of the *Mesoclupea–Paraclupea* assemblage that covers a vast area in southern China, Japan and South Korea. The general morphology of chuhsiungichtyids is similar to that of certain basal osteoglossomorphs, especially the genus *Paralycoptera*, and these fishes must certainly have a very close autoecology.

The cladocyclids are a family defined by Maisey [MAI 91] that contains, according to Cavin *et al.* [CAV 13b], the sister genera *Cladocyclus* and *Chiromystus*, to which should be added *Aidachar*, a genus initially attributed to a pterosaur from the Turonian of Uzbekistan, which was identified as a cladocyclid by Mkhitaryan and Averianov [MKH 11]. *Chiromystus* was found in continental sites in Brazil and southwestern Africa dating from the Hauterivian to the Aptian and represents an example of vicariances associated with the opening of the South Atlantic. *Cladocyclus* is present in the Crato [LEA 04] and Santana Formations in Brazil, as well as various other continental formations in Brazil. This genus can be considered euryhaline. *Aidachar* is present in the Cenomanian of the Kem Kem beds in Morocco [FOR 07], considered to be a deltaic continental environment, and in continental or brackish Turonian deposits of Uzbekistan. The probable euryhaline lifestyle of the cladocyclids makes it difficult to conduct a paleobiogeographical analysis of the distribution of their occurrences (Figure 4.2).

4.13. Elopomorphs (Elopomorpha)

Today, elopomorphs are primarily marine. They are distinguished by their extremely thin, leaf-shaped larvae, known as leptocephalus larvae. Transitions toward continental environments, likely independent of one another, occurred during the Cretaceous with an elopiform from El Montsech in Spain, *Ichthyemidion* [POY 95], and two megalopiforms, *Arratiaelops* and *Paratarpon*, from the Early Cretaceous of Europe (Isle of Wight and Bernissart) and from the Campanian Oldman Formation in Canada, respectively.

Phyllodontids are a mysterious, extinct family, known mainly from isolated dentitions. In these fishes, the teeth are located in the palate (parasphenoid) and the hyoido-branchial apparatus. The teeth, generally circular and thin, are arranged in adjacent columns that together form a kind of dental plate. Phyllodontids are included in the order albuliforms, of which the modern genus *Albula* also has small flat and circular teeth, although not stacked as in the extinct family. In fact, the term "phyllodontid" is more general and covers this particular dental arrangement that can be found in several modern families of fishes, such as labrids, odacids, scarids within the labroids, sciaenids, carangids and diodontids [EST 69]. The monophyly and affinities of the phyllodontids, considered here to be elopomorphs, remains questionable. The phyllodontids are known from the Early Cretaceous to the

Paleogene of North America, Europe and Africa. It seems that we can observe a transition between the oldest forms, *Casierus* and *Paralbula*, which were euryhaline, to more recent forms, *Phyllodus* and *Egertonia* that were strictly freshwater. However, this distribution could also correspond to a diadromous lifestyle, comparable to that of the eel (also an elopomorph). The freshwater forms of elopomorphs in the Cretaceous (*Ichthyemidion*, *Arratiaelops* and *Paratarpon*) could also have been diadromous species.

4.14. Osteoglossomorphs (Osteoglossomorpha)

Osteoglossomorphs, or bonytongues according to the etymology of taxon's name, form a well-defined clade characterized by their unpaired jaw located in the mouth (as opposed to the "pharyngeal jaws" of the cypriniforms) made up of dentigerous plates located on the parasphenoid dorsally and on the basihyal ventrally. Today, osteoglossomorphs are exclusively freshwater, and the same is true throughout a large part of their evolutionary history. However, some returns to a marine environment occurred in the Paleogene [BON 08, FOR 10]. Today, the vast majority of freshwater actinopterygians are otophysans, a group whose evolutionary history took place primarily in the Cenozoic (cf. 4.15.2.2). Osteoglossomorphs, on the other hand, are not very diverse today, and the clade detached from the rest of the actinopterygians (the clupeocephala) well before the otophysans individualized. A large part of their evolutionary history took place during the Mesozoic. It is logical to assume that this history, contemporary to the first phases of the fragmentation of Pangaea, would be strongly marked by vicariant events, or at least, much more than for the otophysans. Various authors have approached the biogeographical history of the osteoglossomorphs on the basis of the fossil record [BON 96, WIL 08, FOR 10] or on the basis of topologies provided by molecular phylogenetic analyses [KUM 00, LAV 04], or by synthesizing these two data sources [LAV 16]. This last study combined molecular (mitochondrial and nuclear) and morphological data of modern and fossil forms to test three vicariance hypotheses for explaining the current distribution of osteoglossomorphs, namely (1) the separation of African and Asian notopterids as it relates to the separation of Africa from the India-Madagascar block 140 million years ago; (2) the separation of modern genera of *Arapaima* in South America and *Heterotis* in Africa as it relates to the opening of the South Atlantic 110 million years ago; and (3) the separation of the modern genera *Osteoglossum* in South America and *Scleropages* in Australasia as it relates to the separation of the Antarctic–South America block from Australia. The first two vicariance hypotheses should be rejected unless we consider the origin age of the crown group of teleosteans to be around 300 million years, which is extremely improbable. The third hypothesis should not be rejected if we consider a reasonable origin age of about 200 million years for the crown group of teleosteans. However, another study by the same author [LAV 15] shows that the genus *Scleropages* carried out a marine dispersal between Australia, Papua New Guinea and southeast Asia, a result that weakens the vicariance explanation for the origin of the genus. Lavoué's model [LAV 16], which involves dispersals rather than vicariances to explain modern distributions and proposes new phylogenetic positions for certain fossils as a result of the combined analysis of molecular and morphological data, will serve as a framework for the discussion that follows. This model also rejects the hypothesis that suggested that India played the role of a "transport raft" as it moved between Gondwana and Asia for the osteoglossins [KUM 00] and notopterids [INO 09] (cf. 5.1).

As for other clades discussed in this book, the topology of the phylogenetic tree plays a fundamental role in the discussion of the biogeographical scenario. A consensus seems to have emerged for the main lines of the phylogeny of the osteoglossomorphs and is compatible with molecular and morphological data (cf. [LAV 16] and [WIL 08]). Considering the modern taxa, the schema is as follows: the osteoglossoids include the osteoglossids with two subfamilies, the osteoglossins (*Osteoglossum* and *Scleropages*) and the heterotins (*Arapaima* and *Heterotis*). The sister group of the osteoglossoids is the notopteroids, which include the notopterids and the mormyroidea, which include the mormyrids and the gymnarchids. Together, osteoglossoids and notopteroids make up the osteoglossiforms, whose sister group is the hiodontiforms. The monogeneric family of the pantodontids, missing from this schema, was placed within the osteoglossids [TAV 98, HIL 03], or as sister group of the osteoglossids [GRE 73, BON 96, WIL 08], or as sister group of the osteoglossiforms [LAV 04, LAV 16]. How are the fossils included in this schema? Do they clarify or confuse it? The oldest known osteoglossomorphs are forms from the continental Early Cretaceous in eastern Asia (Figure 4.5). The data on these taxa can be found in [ZHA 98, LI 99, CHA 04, WIL 08, XU 09, FOR 10], among others. According to studies, these genera belong in part to the extinct lycopterids (*Lycoptera*) and to the modern hiodontids (*Jiaoichthys*), are in basal position of crown osteoglossomorphs (*Paralycoptera, Tanolepis, Xixiaichthys, Tongxinichthys*) and in basal position of crown osteoglossiforms

(*Kuntulunia, Huashia*). The phylogenetic position of these genera is fairly liable to change from one study to another. If this general topology is confirmed, in particular the positions of *Kuntulunia* and *Huashia* as basal osteoglossiforms [ZHA 98, ZHA 06] or even as "arapaimines" [LI 99], then we can consider that the first part of the evolutionary history of osteoglossomorphs occurred in eastern Asia. Forey and Hilton [FOR 10], however, are skeptical of the connection of *Kuntulunia* and *Huashia* to a subgroup of osteoglossomorphs and even of their inclusion within the osteoglossomorphs at all. The phylogenetic position of the basal osteoglossiform *Tetoriichthys* in Japan varies according to the matrix of characters used: it appears to be either an osteoglossin, the sister genus of the notopterids or a basal osteoglossiform [YAB 08]. This taxon, ancient for an osteoglossomorph, reinforces the idea of an Asian origin for the group. The hiodontids persisted in Asia at least until the Campanian as indicated by the presence of isolated vertebrae in the Nemegt Formation in Mongolia [NEW 13]. Quickly, like several other groups during the Cretaceous, dispersals occurred from Asia toward North America where hiodontids still survive today endemically. The fossil record of hiodontids in North America begins in the Cenomanian of Utah [BRI 13] with the presence of isolated elements attributed to this family. In the Cenozoic, their fossil record is composed exclusively of the genus *Hiodon*, present since the Eocene, since the genus *Eohiodon* was synonymized with the modern genus *Hiodon* [HIL 08]. Notably, it is also in North America, in the Paleocene of Montana, that *Ostariostoma* was described, a genus considered in recent studies to be the sister taxon of the osteoglossiforms ([HIL 03, WIL 08, LAV 16], but see [ZHA 06] for an alternative solution). The presence of stem osteoglossomorphs in North America was very recently confirmed with the discovery of *Wilsonichthys* in the Maastrichtian of Alberta, Canada [MUR 16]. It is closely related to *Shuleichthys* from the Early Cretaceous of China described a few years earlier [MUR 10b]. These data tell us that the osteoglossiforms appeared in Asia and/or North America during the Cretaceous. Some of these basal taxa survived until the Paleogene of North America. In addition, two species attributed to the genus *Joffrichthys* were recognized in the North American Paleocene. This genus is considered to be a heterotin by Li and Wilson [LI 96] and Wilson and Murray [WIL 08], and to be a basal osteoglossid by Hilton [HIL 03] and Lavoué [LAV 16].

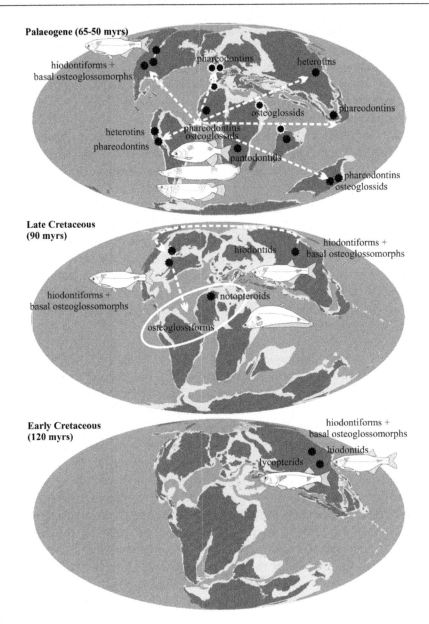

Figure 4.5. *Evolutionary history of osteoglossomorphs during the Mesozoic and Paleogene. Vicariances are represented by double-ended arrows, dispersals by dotted arrows and occurrences by stars (stars with a white border are marine occurrences). This sequence represents one possible scenario. Another scenario is depicted in Figure 4.7. For a color version of the figure, see www.iste.co.uk/cavin/fishes.zip*

According to Lavoué [LAV 16], the pantodontids are the most basal modern osteoglossiforms. The strictly African distribution of this family implies *a priori* a dispersal from Laurasia toward Africa in the Late Cretaceous or at the very start of the Paleogene. Two continental genera from the Eocene of Mahenge in Tanzania, *Singida* and *Chauliopareion*, were included in the pantodontids or were connected to this family [MUR 05, WIL 08, LAV 16]. One potential candidate that illustrates a marine phase indicative of dispersal from the northern hemisphere toward Africa in the Cretaceous is *Prognathoglossum* from the marine Cenomanian of Lebanon, which was described as a pantodontid by Taverne and Capasso [TAV 12]. However, although some very derived cranial characters do indeed recall the genus *Pantodon*, the post-cranial skeleton is very different. A cladistic analysis including *Prognathoglossum* in a large sample of osteoglossomorphs, notably *Singida* and *Chauliopareion*, is needed to test the belonging of this strange genus to the pantodontids. Considering the Asian and/or North American origin for osteoglossomorphs, an early marine dispersal is also necessary to explain the presence of notopteroids in western Gondwana in the "mid" Cretaceous (see below).

Several extinct osteoglossomorphs were described from marine environments and are worth mentioning in the biogeographical history of this group. Some of these genera, however, have a doubtful attribution to the osteoglossomorphs according to Forey and Hilton [FOR 10]. This includes *Foreyichthys*, *Thrissopterus* and *Monopterus* from the Eocene of Monte Bolca in Italy. Other marine osteoglossomorphs were recently described from the Eocene of Denmark, including *Furichthys*, *Brychaetoides*, *Xosteoglossid* and *Heteroglossum*, and according to Bonde [BON 08], they all correspond to independent returns to marine environments. Another well-defined group that contains a marine genus alongside freshwater forms is the phareodontins, a subfamily positioned as a sister group to the osteoglossins [LAV 16] or to the heterotins [WIL 08]. The marine genus is *Brychaetus muelleri* from the Eocene of the Isle of Sheppey in England, also found in the Eocene phosphate deposits of Morocco and in Rájasthán, India [FOR 10]. The most ancient representatives of this subfamily were found in the continental Campanian of Alberta, Canada. This genus, *Cretophareodus* is considered to be close to *Phareodus* by Li [LI 96] and to be an osteoglossid by Wilson and Murray [WIL 08]. The other freshwater forms of this clade are *Phareodus*, a well-known genus with two species, *Phareodus encaustus* and *P. testis*, both from the Eocene of the Green River Formation and *P. queenslandicus* from the Eocene of Queensland [LI 97], as well as

Phareodusichthys and *Taverneichthys*, two much less well-known genera from the Paleocene of Bolivia [GAY 98] and India [KUM 05], respectively. The presence of the genus *Phareodus* in the continental waters of North America and Australia during the Eocene might be surprising, but Li *et al.* [LI 97] found that the Australian species together with one of the American species (*P. encaustus*) form a clade that is a sister group to the marine form *Brychaetus*. Consequently, we can suppose that the continental forms of phareodontins were "captured" from marine environments rather than being the result of vicariance events. The identification of *Chanopsis*, from the Aptian of the Democratic Republic of Congo (formerly Zaire) as a phareodontid by Taverne [TAV 84] was challenged by Forey and Hilton [FOR 10], who doubted that it belongs to the osteoglossids. Another marine genus from the Paleocene of Saudi Arabia, *Magnigena*, is closer to the osteoglossins than to the phareodontins. Finally, a piece of a skull from an unnamed taxon from the marine Eocene of London Clay, in England, is considered to be very close to *Scleropages* and *Osteoglossum* [FOR 10]. The existence of various marine forms among basal osteoglossiforms makes it likely that marine dispersals occurred in the Cretaceous. It notably explains the Gondwanan distribution of osteoglossiforms in relation to the Laurasian basal osteoglossomorphs (Figure 4.5).

Isolated remains attributed with uncertainty to *Scleropages* were recorded in the Campanian of France [SIG 97, TAV 09], in the Maastrichtian of India [KUM 05], in the Paleocene of Niger [TAV 09] and Belgium [TAV 07], in the Eocene of Sumatra [FOR 10] and in the Oligocene of Australia [UNM 01]. However, Lavoué [LAV 15] noted that, in the absence of a synapomorphy making it possible to distinguish *Scleropages* from *Osteoglossum*, these elements must be considered to belong to indeterminate osteoglossins instead. In this study, Lavoué rejects the idea of a vicariance between *Osteoglossum* and *Scleropages* during the separation of Gondwana based on a chronological argument. However, he also rejects the possibility of a marine dispersal, leaving open the question of the geographical origin of the genus *Scleropages* in Asia. The fossil record of the other subfamily of osteoglossids, the heterotins, starts with the fossil of a taxon close to *Arapaima* dating from the Paleocene of Bolivia [GAY 98, FOR 10]. Then, *Sinoglossus* from the Eocene of China is evidence of the presence of the *Arapaima* lineage in Asia. This occurrence substantially, as well as surprisingly, increases the distribution area of this lineage. *Heterotis* is known in Africa since the Early Miocene [STE 01]. The fossil record of the notopteroids is sparse. The only presence recorded in the Cretaceous is

Palaeonotopterus from the Cenomanian of the Moroccan Kem Kem beds. Initially identified as a notopterid [FOR 97], its position has since varied: as a sister group to the mormyroids [WIL 08] or with uncertain relationships within the notopteroids [CAV 01] or within the osteoglossiforms [HIL 03]. New, more complete specimens in the process of being studied should make it possible to resolve the phylogenetic position of this particular genus, which is currently known primarily from braincases and a few associated elements.

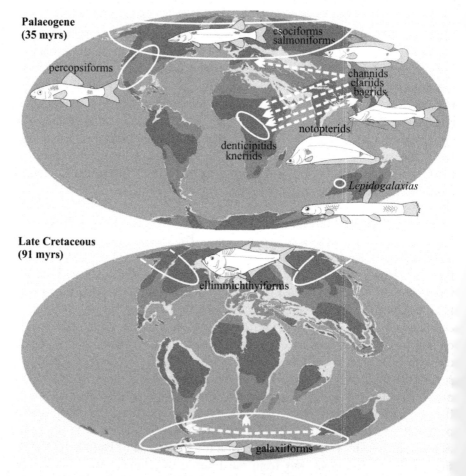

Figure 4.6. *Distribution areas of freshwater clades during the Late Cretaceous and the Paleogene. The green zones correspond to humid zones, where precipitation was greater than evaporation (according to Scotese [SCO 14]). For a color version of the figure, see www.iste.co.uk/cavin/fishes.zip*

The fossil record of crown osteoglossiforms provides a relatively confusing biogeographical history. As discussed, it cannot be explained exclusively by vicariance events related to continental fragmentation. The current geographical distribution is mainly the result of dispersals that are difficult to read due to the existence of marine taxa. The latter are mainly osteoglossiforms from the Paleogene, a period when the fossil record increased the geographical distribution of the group considerably. We can assume that the dispersals at the origin of the pantodontid and notopteroid lineages toward Africa during the "mid" Cretaceous, and then the dispersals of the osteoglossoids toward Africa and Eurasia in the terminal Cretaceous, occurred from west to east. The evolution of notopteroids in Africa, with the development of a substantial evolutionary radiation of mormyrids, only left fossils from the Late Miocene onward [STE 01]. However, the mormyrids must be much more ancient than that, because the sister family, the gymnarchids, is known in North Africa from the Eocene on. We can assume that the Asian notopterids resulted from a dispersal when the African continent made contact with Eurasia in the Miocene [OTE 10] at a time when the group was already living exclusively in freshwater. On a map of the Oligocene that shows rainy climatic zones, an environmental continuity can be observed between central Africa and southern Asia that would have facilitated this dispersal (Figure 4.6).

The scenario discussed above involves the first part of the history of the osteoglossomorphs in Asia and/or North America and is entirely based on the phylogenetic position of several genera located in a pectinate position at the base of the clade. Cladograms, based on morphological data and which also contain extinct taxa, often have weakly supported branches (small changes in the coding of characters can easily modify topologies). The trees often go through significant transformations when new taxa or new characters are discovered. Based on the fragility of these data, we can propose the following suboptimal alternative hypothesis to explain the recent distribution of osteoglossomorphs. In this scenario, all the basal osteoglossomorphs from the Cretaceous in Asia (lycopterids, hiodontids and a few other genera), and from the Cretaceous to the current period in North America (hiodontids, *Wilsonichthys*, *Ostariostoma*) would form a clade. In this case, the fundamental structure of the osteoglossomorph tree could be explained by an initial vicariance between Laurasia and Gondwana in the Late Jurassic or the Early Cretaceous [CAV 08]. Then, the dispersal events described above would have followed this initial split (Figure 4.7). This initial vicariance would be comparable, and even contemporaneous, to the vicariance between the cypriniforms in Laurasia and the characiphysi in Gondwana. This scenario seems simpler than the previous one but it it is difficult to explain the presence of basal osteoglossiforms in Asia and North America during the Cretaceous and the Paleogene. Insofar as the phylogenies of the osteoglossomorphs present important differences according to the authors, it is not unlikely that this scenario will be validated in the future.

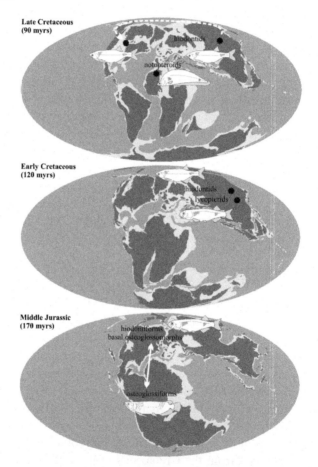

Figure 4.7. *Evolutionary history of osteoglossomorphs during the Mesozoic era. Vicariances are represented by double-ended arrows, dispersals by dotted arrows and occurrences by stars (stars with a white border are marine occurrences). This sequence represents an alternative scenario to the one depicted in Figure 4.5. In this scenario, it is difficult to account for basal osteoglossiforms in Asia (Kuntulunia, Huashia and Shuleichthys) and North America (Ostariosoma and Wilsonichthys). For a color version of the figure, see www.iste.co.uk/cavin/fishes.zip*

4.15. Otomorphs (Otomorpha)

Recently, the clupeomorphs were grouped with the gonorynchiforms and the otophysans within the otomorphs (Otomorpha according to Betancur *et al.* [BET 13]).

4.15.1. *Clupeomorphs (Clupeomorpha)*

Clupeomorphs are a diversified group that is distinguished by a particular connection between the air bladder and the inner ear by two diverticula extending from the air bladder and penetrating through the back of the braincase where they meet the inner ear. Another feature, generally more visible on fossils, is the presence of scutes, which are large, club-shaped scales aligned along the ventral axis of the body. In certain modern species, but especially in fossil species, scutes may be present along the median dorsal line in front of the dorsal fin, and even in front and behind this fin, as is the case with the appropriately named *Triplomystus*. Today, clupeomorphs contain the only clupeiforms, but they are related to the ellimmichthyiforms, an order that went extinct in the Eocene. The two orders are primarily marine, but they show several transfers toward freshwater environments with either diadromous species or species blocked into continental environments. Based on a molecular analysis and by reconstructing ancestral environments, Lavoué *et al.* [LAV 13] detected in the clupeoids at least 11 transfers from marine to freshwater environments during the Cenozoic. Very few of these continental forms left fossil traces. The second order of the clupeomorphs, the ellimmichthyiforms, left a richer fossil record in both marine and continental deposits. In this order, also, there seem to have been several transfers between these two environments, as can be observed if the living environments are positioned in a phylogeny of the clade, such as the one by Murray and Wilson [MUR 13]. This means that since its beginning, probably in the Late Jurassic, this group has been characterized by species with a physiology that allowed them to adapt to different environments. This characteristic makes it impossible to recognize clear paleobiogeographical schemas by only reading the fossil record. More cautiously, we can point out some continental paleontological deposits that yielded clupeomorphs and attempt to draw some general conclusions from these observations. Various localities in the Late Cretaceous in South America contain *Gasteroclupea branisai*, a species related to modern pristigasteroidea according to Grande [GRA 82] and to the pristigasterins according to Arratia and Cione

[ARR 96]. Deposits from the Early Cretaceous of eastern Asia contain continental derived ellimmichthyiforms, such as *Paraclupea* in China [CHA 97] and *"Diplomystus"* in Japan [YAB 94] (Figure 4.6). The genus *Diplomystus*, a mostly marine genus, is present in North American freshwater since the Cenomanian of the Dakota Formation and the Early Santonian of the Straight Cliffs Formation, as well as from the Campanian of the Dinosaur Park Formation [DIV 16]. It has also been found in various sites in the Eocene of North America, including the Green River Formation, and it disappears at the end of this period. In the Paleocene, the clupeomorph *Primisardinella* is described from Turkistan [GRA 85, MUR 05]. In North American, the freshwater genus *Knightia* is present since the Paleocene of the lacustrine deposits of Montana [GRA 85] and has been found in large numbers in the Eocene of the Green River Formation. In western Africa, the Early Cretaceous of Benito in Equatorial Guinea yielded clupeomorphs whose complex taxonomic history is summarized by Murray *et al.* [MUR 05]. Freshwater clupeids are known in Africa in the Cenomanian of the Djebel Oum Tkout site in southeastern Morocco [DUT 99], and later in the Oligocene of Fayoum with *Chasmoclupea* [MUR 05]. The extant monospecific family of the denticipitids, the most basal living clupeomorphs, is endemic to a few coastal rivers in eastern Africa. *Denticeps clupeoides*, which was only discovered in 1959, is distinguished by odontodes, a kind of teeth borne on the surface of its cranial bones. *Palaeodenticeps*, from the Eocene of Mahenge in Tanzania, is the most ancient representative of this family [GRE 60].

4.15.2. Ostariophysans (Ostariophysi)

Gonorynchiforms are grouped with otophysans within the ostariophysans (Ostariophysi), which are themselves grouped with the clupeomorphs within the otomorphs.

4.15.2.1. Gonorynchiforms (Anatophysa)

The order gonorynchiforms includes two families, the chanids and the gonorynchids. Today, the gonorynchids are only represented by a single genus and five marine species. The family is known since the base of the Late Cretaceous. It is essentially marine with the exception of *Notogoneus*, which is present in the Cretaceous and Paleogene. This genus, whose fossil record was detailed by Grande and Grande [GRA 99], appears in the Campanian of the Two Medicine Formation in Montana, USA. During the Cenozoic, they are found in the Eocene of the Green River Formation in the United States, in the Late Eocene or Oligocene of Queensland in Australia, in the Eocene of Ukraine, in the Late Oligocene or the Early Miocene of Germany and in the Oligocene of various French sites. According to Grande and Grande [GRA 99], this very large geographical distribution is linked to a diadromous lifestyle. The diadromous lifestyle was suggested because of the distribution of body sizes in a sample of well-preserved specimen of a population of *Notogoneus osculus* from the Eocene of the United States, which is characteristic of a population in the midst of migration [GRA 99]. To my knowledge, this is one of only a few examples of paleontological evidence of a diadromous lifestyle for an extinct fish. Another study indicating anadromy [CAR 03] is based on an isotope analysis of otoliths from a possible ostariophysan from the Maastrichtian of the United States that demonstrated an anadromous lifestyle (life at sea and reproduction in brackish water).

The other clade of gonorynchiforms is made up of chanoids (Chanoidei). Included in this clade are kneriids and phractolaemids (sometimes considered to be subfamilies), freshwater fishes that are endemic to the African continent. Today, kneriids include five genera of small fishes that live high up in rivers. Two of these, *Cromeria* and *Grasseichthys*, are pedomorphic, which means that the adults retain juvenile traits. To my knowledge, the only known representative fossil is *Mahengichthys singidaensis* from the Middle Eocene of Mahenge in Tanzania [DAV 13]. The discovery of this species, which is not pedomorphic, suggests that the family individualized in sub-Saharan Africa and remained there. Phractolaemids only include one species, *Phractolaemus ansorgii*, which is a small fish with a very specialized jaw. We have no fossils from this family. Within the chanoids, the chanids are only currently represented by the species *Chanos chanos*, which is marine, but whose juveniles can enter rivers and lakes. The fossil record of chanids, which is richer than the other clades of gonorynchiforms, begins in the Early Cretaceous [GRA 99, POY 10, FAR 07, FAR 10]. Certain genera from the Early Cretaceous come from continental or lagoonal paleoenvironments, such as *Aethalionopsis* from the Wealden in Bernissart and *Parachanos* from the Aptian of Cocobeach in Gabon, or from closed marine environments like *Dastilbe* and *Tharrhias*, which are from the Aptian of the Crato Formation and the Albian of the Santana Formation in Brazil, respectively. *Parachanos* and *Dastilbe* are phylogenetically close and the latter, which is very common in the Crato Formation, may even be present on the African side of the Atlantic, in Equatorial Guinea [GAY 89]. *Dastilbe* probably lived in freshwater or was anadromous [DAV 99]. Although the alpha-taxonomy of the *Dastilbe* and *Parachanos* genera on both sides of the South Atlantic requires clarification, it appears that vicariance phenomena at the genus or species level took place within the clade during the Early Cretaceous (Figure 4.2). Other freshwater chanids are present in the Early Cretaceous in Spain, with *Rubiesichthys* and *Gordichthys* in the Barremian of Las Hoyas and *Rubiesichthys* in the Berrasian of El Montsech. These two are sister genera [POY 10], to which the closest taxon is *Nanaichthys* in the Aptian of the Marizal Formation in northeastern Brazil [AMA 12]. These authors suggest a dispersal event between Europe and South America through the western Tethys as it opened. The most recent phylogeny of the chanids [AMA 12] indicates that the modern *Chanos*, a marine genus, is nested deeply within the family, while the more basal members come from freshwater or brackish environments in the Cretaceous. This situation contrasts with the majority of Mesozoic clades that have freshwater and marine representatives, but whose current forms are confined to freshwater environments (refuge environments).

4.15.2.2. *Otophysans (Otophysi)*

Otophysans (cypriniforms, characiforms, siluriforms and gymnotiforms) represent 68% of the current diversity of freshwater fishes. Siluriforms themselves represent 5% of all vertebrates. The most distinguishing feature of the group is the Weberian apparatus, a set of substantial modifications in the first few vertebrae and their associated elements that makes it possible to transmit vibrations between the air bladder and the inner ear. It is essentially a median vertebral ear in which the ossicles of the middle ear of mammals are replaced by elements of the vertebrae. The paleontological origins of otophysans are still not well understood. Patterson [PAT 75] discussed this origin and concluded in a rather forward-thinking way. He observed that within the gonorynchiforms, the sister group of the otophysans, we know of a representative of the group from the Early Cretaceous, *Tharrhias araripis*. In this species, the synapomorphies of the clade are clearly present in the skull, but they are largely absent from the caudal skeleton. Applying this observation to the otophysans, he pointed out that the fossil species could very well have a Weberian apparatus without actually possessing the synapomorphies of the clade in the caudal skeleton. Knowing that the first vertebrae are often hidden by the opercular series in the fossils, these forms can, according to Patterson, pass unnoticed by paleontologists. He suggested certain little-known species belonging to the ill-defined genera of *Clupavus* and *Leptolepis* as potential, non-identified candidates for basal otophysans.

And, indeed, modified anterior vertebrae have since been observed on a form close to *Clupavus*, *Lusitanichthys characiformis*, from the Cenomanian of Portugal [GAY 81], and later on a species from the genus *Clupavus*, *Clupavus maroccanus* from the Cenomanian of Morocco by Taverne [TAV 95], and more recently, on *Lusitanichthys africanus* from the Cenomanian of Morocco [CAV 99a]. Gayet [GAY 81, GAY 85] and Taverne [TAV 95] have detected features in the two genera that indicate affinities with the characiforms for *Lusitanichthys* according to Gayet, and with the characiforms plus the siluriforms for *Clupavus* according to the Taverne. Cavin [CAV 99a] challenged the affinities at level of the order and only considered that the two genera probably belong to basal otophysans. While Gayet and Taverne observed a rudimentary Weberian apparatus in two genera dating from the base of the Late Cretaceous, Patterson [PAT 84] also described a Weberian apparatus in *Chanoides macropoma* from the marine Eocene of Monte Bolca in Italy (Figure 1.3). He also noted some other otophysan features present at the end of the snout that prompted him to consider this species to be a stem otophysan. In the genus *Chanoides*, Taverne [TAV 05] placed a second species, *Chanoides chardoni* from the Campano-Maastrichtian of Nardo in Italy. It is now included in the genus *Nardonoides* and is placed at the base of the ostariophysans (gonorynchiforms and otophysans) according to Mayrinck *et al.* [MAY 15]. Finally, *Santanichthys diasii*, from the Santana Formation in Brazil, is also considered to be a basal characiform by Filleul and Maisey [FIL 04]. On the other hand, Malabarba and Malabarba [MAL 10] consider this form to be a basal ostariophysan or a basal otophysan. An interesting point that comes out of this review is that all of these potential basal otophysans come from marine or brackish paleoenvironments. The otophysan clade therefore cannot be considered to be archeolimnic.

Based on a new molecular phylogeny and a new chronology, Chen *et al.* [CHE 13] suggest an interesting scenario that describes the origin and spatial distribution of different orders of otophysans (Figure 4.8). In this model, the separation between the cypriniforms and all other otophysans (the characiphysi) is the result of a vicariance dating from the Late Jurassic, about 150 million years ago, that separated the cypriniforms in Laurasia from the characiphysi in Gondwana. Within the Gondwanan clade, on the western side corresponding to South America, the gymnotiforms detached from the other characiphysi at the start of the Cretaceous. Within the characiphysi, according to this model, the characiforms are no longer monophyletic but the citharinoidei (an African clade generally considered to be a sister group of other characiforms) appeared as the sister group of the remainder of the characiforms plus the siluriforms. This topology was also found by Nakatani *et al.* [NAK 11]. Consequently, a vicariance occurred between the citharinoidei and the other characiforms plus siluriforms during the opening of the South Atlantic that separated South America from South Africa during the Early Cretaceous. The separation of characiforms *sensu stricto* and siluriforms took place within the South American continent during the Early Cretaceous. The current geographical distribution of siluroidei and cypriniforms, which are both ubiquitous with the exception of the absence of cypriniforms in South America and Australia, was considered by Chen *et al.* [CHE 13] to be the result of dispersals during the Cenozoic. This model seems much more probable than the one proposed by Nakatani *et al.* [NAK 11] which, based on a relatively similar phylogenetic analysis of mitochondrial genes, suggested that the origin of the otophysans dates back to the Permian and the origin of the characiphysi dates back to the Triassic. These dates are definitely much too old compared to the fossil record. In the present work, the model of Chen *et al.* [CHE 13] serves as a template on which fossils are positioned. The issue now, already raised by Chen *et al.*, is that the initial separation of the cypriniforms from the characiphysi seems to have occurred between western Gondwana (South America and Africa) and southeastern Asia, where cypriniforms have supposedly originated, two geographic domains that never touched.

4.15.2.2.1. Cypriniforms

This very diverse order is common in freshwater environments in the northern hemisphere and Africa, and absent from South America and Australia. They are primarily distinguished by the presence of an extra bone in their snout called the kinethmoid, the absence of buccal teeth and the presence of pharyngeal teeth. These can be common in certain Cenozoic fossiliferous sites and are, to a certain extent, diagnostic. The fossil record of cyprinids for this period was exhaustively presented by Cavender in 1998 [CAV 98]. The occurrences are not detailed here if they do not play a role in understanding the biogeographical history of the clade. Some molecular data indicate an origin of the cypriniforms in southeastern Asia toward the end of the Triassic, followed by a diversification and the appearance of the crown group in the Late Jurassic [SAI 11]. These data contradict our knowledge of faunas for these periods and regions. Information about the fossil record in question, i.e. fossil fishes from the continental Mesozoic in southeastern Asia, was very fragmentary only two decades ago. However, it has progressed in recent years because of work conducted in northeastern Thailand. This work has yielded a wide variety of freshwater ginglymodians (cf. 2.3.3.2) and, for the moment, a total absence of ostariophysans and more generally, teleosteans [DEE 16]. An interesting observation by Saitoh *et al.* [SAI 11] about the absence of cypriniform fossils in the Mesozoic brought up the fact that if these forms were fluvial, like most of them are today, the likelihood of their fossilization is low. However, in my opinion, this argument seems insufficient to explain the almost complete absence of forms other than ginglymodians in the Late Jurassic and the Early Cretaceous in Thailand, even in sites that preserve small isolated elements.

An alternative hypothesis to a southeastern Asia origin is proposed here. It is based on the mostly North American distribution of the most basal clade of cypriniforms, the catostomids. According to this scenario (Figure 4.9), a vicariance occurred between North and South America about 150 million years ago, and the cypriniforms then quickly dispersed through Beringia toward Asia where they diversified. This scenario, involving the passage of cypriniforms through North America to get to Asia, was challenged following the discovery of the extant *Paedocypris* in southeastern Asia [KOT 06]. This fish, the smallest known living vertebrate, was placed as the sister group to all other cypriniforms by Mayden and Chen [MAY 10] based on a molecular analysis. This phylogenetic position supports the order's origin in southeastern Asia. However, in a critical review of Mayden and Chen's phylogenetic analysis completed by a morphological phylogeny, Britz *et al.* [BRI 14] challenged the basal position of *Paedocypris* within the cypriniforms. In this study, the catostomids returned to their basal position that they previously occupied. Thus, catostomids are considered here to be the sister group of all other cypriniforms, although alternative solutions place loaches (cobitoidea) as the sister group to all other cypriniforms, including the catostomids [SAI 11], or the catostomids as the sister group to loaches, with all of them forming the sister clade of cyprinoids [MAY 10]. Today, the catostomids have a fragmented distribution area. They are diversified in North America, with a species common to Alaska and the far east of Siberia, and a single isolated species in southern China. The fossil record of catostomids is relatively plentiful in the Eocene-Oligocene of North America, in the United States and Canada, in particular with the genus *Amyzon* [WIL 77, GRA 82], which is probably present from the Paleocene onwards, with its first in the Paskapoo Formation in Canada [WIL 80]. The occurrences are equally numerous in the Neogene [SMI 81, SMI 13]. The catostomids are also present in the Chinese fossil record since the Eocene where the genus *Amyzon* was also recognized in a large geographical area [CHA 01, CHA 08], thus filling the gap in the current distribution of the family. In China, the fossils of this genus were long placed in the modern genus *Osteochilus*, a cyprinid, before being attributed to the catostomid genus *Amyzon*. Recently, a genus close to the extant Asian genus *Myxocyprinus*, *Plesiomyxocyprinus*, was described in the Middle Eocene of northeastern China, implying an ancient divergence time between the Asian and North American taxa [LIU 09].

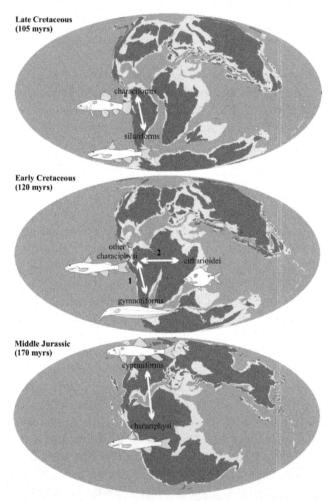

Figure 4.8. *Evolutionary history of orders of ostariophysans during the Mesozoic. Vicariances are represented by double-ended arrows. These different steps are not illustrated by fossils. For a color version of the figure, see www.iste.co.uk/cavin/fishes.zip*

The sister clade of all cypriniforms except the catostomids is made up of cobitoids currently present in Eurasia (as well as Morocco). Fossils of cobitids are present in the Miocene of China and central Asia [CHA 08]. Then, situated in a more derived position in relation to the cobitoids in the phylogeny by Chen *et al.* [CHE 13], a series of little-diversified families (psilorhynchids, danionids, sundadanionids, leptobarbids) are endemic to southeastern Asia. They attest to a radiation in this region at the end of the Cretaceous or the start of the Tertiary according to the phylogeny by Chen *et al.* [CHE 13]. The most derived and the most diversified family is that of the cyprinids. The oldest cyprinid fossils in Europe and North America are from the Late Eocene or the Early Oligocene of the Cypress Hills Formation [DIV 15] in Saskatchewan and in the Early Oligocene of Switzerland [GAU 77]. The two occurrences are included in the derived leuciscins, with the modern genera aff. *Ptychocheilus* in North America and *Leuciscus* in Europe. The Oligocene of North America yielded some remains of indeterminate cyprinids in the Gumboot Mountain site and the Weaverville Formation [CAV 98]. In the Middle and Late Oligocene, the genera *Tarsichthys*, *Palaeotinca*, *Protothymallus* or *Varhostichthys* appeared [GAU 79, GAU 84a, GAU 13]. The extinct genus *Tharsichthys* is found in the Oligo-Miocene of the Swiss molasse [GAU 02] and in the Late Oligocene of Germany [GAU 07], together with the genus *Palaeorutilus* [GAU 88]. In the Neogene, the European cyprinids essentially belong to the modern genera *Tinca*, *Leuciscus* and *Barbus*, added to which are a few extinct genera such as *Palaeoleuciscus* and *Palaeocarassius*. The fossil record of cypriniforms in China and in neighboring regions was revised by Chang and Chen [CHA 08]. The oldest cyprinid in this region is *Palaeogobio zhongyuanensis* from the Middle Eocene of the Shahejie Formation in Henan and *Rostrogobio maritima* from the Late Eocene–Early Oligocene of the province of Primorye in eastern Siberia. They belong to the gobionins, the sister subfamily of the leuciscins. Gobionins mostly diversified in eastern Asia but a few genera are distributed throughout Eurasia, like the gudgeon (*Gobio gobio*). The barbel (cyprinins) *Parabarbus mynsajensisa* was described in the Middle Oligocene of the Turgai basin in Kazakhstan. The Oligocene of China mainly provided isolated pharyngeal teeth that were attributed to leuciscins and cyprinins. Sytchevskaya ([SYT 86], cited by [CHA 08]) described nine genera of cyprinids in the Oligocene of Kazakhstan, Mongolia and eastern Siberia that she classifies as leuciscins, gobionins and cyprinins, although Chang and Chen [CHA 08] consider this unlikely. Occurrences of cyprinids in eastern Asia are more common in the

Neogene and are detailed by Chang and Chen [CHA 08]. The most ancient African cyprinids date from the Late Miocene of Egypt [PRI 14] and Tunisia [GRE 72] and resulted from a dispersal from central Asia, probably in the Burdigalian through the Arabian plate, as indicated by isolated elements attributed to *Barbus* in Saudi Arabia [OTE 01a, OTE 01b].

According to a phylogenetic analysis by Mayden *et al.* [MAY 09], the most basal subfamilies of cyprinids are confined to the Old World, with varying geographical areas for cyprinins, rasboins, acheilognathins, tincins and gobionins. Gobionins are identified as a sister group of the leuciscins, which are the only one of these families present in North America in addition to their presence in northern Eurasia. According to Tang *et al.* [TAN 11], the most basal clade of the gobionins are found in southeastern Asia. Imoto *et al.* [IMO 13] proposed a molecular phylogeny of the leuciscins accompanied by a biogeographical scenario. They found clearly distinct clades for Europe, North America and Asia, with the most basal clade being European. They suggested that the subfamily has its origin in Europe and dispersed to North America and then Asia through Beringia in the Late Cretaceous. Although the ages obtained by these authors seem to have been overestimated (the origin of cyprinids in the Paleogene obtained by Chen *et al.* [CHE 13] is more acceptable), their dispersal schema will be retained here.

According to this new model (Figure 4.9), the biogeographical history of cypriniforms corresponds to a complete westward spread around the world starting from North America. In summary, after their origin on this continent from a vicariance with the characiphysi in South America, the clade diversified with the catostomids that rapidly dispersed in Asia through Beringia. The clade then diversified in eastern Asia in the form of several families endemic to that region (psilorhynchids, danionids, sundadanionids, leptobarbids). The ancestor of these Asian families, which technically cannot be a catostomid if this family is monophyletic, could have accompanied the "true" cotostomids on their dispersal through Beringia. The cyprinids also originated in this region. They diversified into several subfamilies: cyprinins, rasboins, acheilognathins and tincins, several of which dispersed toward Europe and Africa. The most recent dichotomy at the origin of the gobionins in Asia and the leuciscins in Europe, which probably dates from the first half of the Paleogene, corresponds to a vicariance phenomenon caused by the Turgai Strait that separated Europe and Asia at the time. The leuciscins then dispersed toward North America, then toward eastern Asia again through

Beringia. Through this scenario, one explains the presence of both the most basal family of cypriniforms, the catostomids, and one of the most derived subfamilies, the leuciscins, in both North America and East Asia. This scenario corresponds to data now available from fossils and phylogenies. It will no doubt be subject to substantial modifications, but it may serve as a working hypothesis for future research.

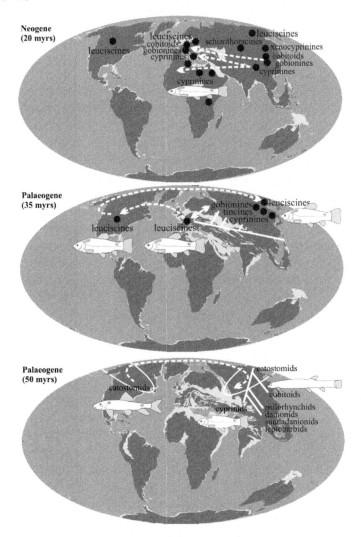

Figure 4.9. *Evolutionary history of cypriniforms during the Cenozoic. Vicariances are represented by double-ended arrows, dispersals by dotted arrows and occurrences by stars. For a color version of the figure, see www.iste.co.uk/cavin/fishes.zip*

4.15.2.2.2. Gymnotiforms (Gymnotiformes)

Gymnotiforms are endemic to the neotropics. They have an easily recognizable morphology and capacities for electroreception and electroproduction (cf. 1.2.4). The order is diverse with more than 150 species. According to a calibrated molecular phylogeny, gymnotiforms detached from other characiphysi in the Early Cretaceous [CHE 13], but the only known fossil is much more recent. This fossil is *Ellisella kirschbaumi*, discovered in the Late Miocene of Bolivia [GAY 91b].

4.15.2.2.3. Characiforms (Characiformes)

Environmentally, the characiforms "replace" the cypriniforms partially in Africa, where the cypriniforms are not very diversified, and completely in South America, where the cypriniforms have always been absent. There are more than 2,000 species of characiforms described so far and they occupy a very wide variety of ecological niches. The current distribution of characiform families, limited to Africa and South America, intuitively suggests an evolutionary history connected with the opening of the South Atlantic. However, the previous distribution of characiforms was broader than just western Gondwana. In addition, the phylogenetic relationships within the characiforms do not fit with a single vicariance event at the origin of the modern distributions. This situation led some researchers to suggest that several extinctions in Africa, where the diversity is lower than in South America, had erased the traces of the vicariances [LUN 93]. Other proposed hypotheses include unbalanced vicariances [MAL 10] or marine dispersals [CHE 13].

The scenario proposed by Malabarba and Malabarba [MAL 10] is based on a model that favors vicariance phenomena and suggests extensive ancestral distributions. It suggests that the diversification of characiforms began before western Gondwana broke away. Because the geographical distribution of these first lineages was not homogenous, the vicariances did not affect all clades in the same way. This hypothesis is supported by the evidence that the opening of the South Atlantic was a complex event that occurred in several steps [MAI 00]. For a time, a part of northeastern Brazil was probably biogeographically connected to Africa while the rest of the South American continent had already separated from the African continent [MAI 11]. This model could support the hypothesis of "unbalanced" vicariances at the origin of the large clades of characiforms. Incidentally, the results of recent molecular analyses indicate an ancient origin (Early

Cretaceous) for the first dichotomies of characiforms [ARR 13], and in particular, the citharinoidei. This scenario cannot be proved for the time being, but it could be tested in the future with the discovery of new fossils on both sides of the South Atlantic.

The fossil record of the characiforms was reviewed and commented on by Malabarba and Malabarba [MAL 10]. It should be noted that some articulated specimens from the Mesozoic that were previous attributed to the characiforms, such as *Lusitanichthys characiformis* in Portugal, *Santanichthys diasii* in Brazil and possibly *Sorbinicharax verraesi* in Italy are now regarded as basal ostariophysans instead (cf. 4.15.2.2). In the Cenomanian of the Kem Kem beds in Morocco, Dutheil [DUT 99] reported, without figurations, two teeth and some vertebrae belonging to characiforms. Prior to that, Werner [WER 94] had attributed three teeth from the Cenomanian site of Wadi Milk in Sudan to a characiform. Werner's illustration of one of the incomplete teeth shows a fragment of a crown with three broken cuspids. This morphology could be connected to multicuspid teeth discovered in association with pharyngeal teeth in various Cretaceous sites in Africa, India and China, and which have been cautiously attributed to a ginglymodian [MO 16]. The oldest definite African characiforms are *Eocitharinus macrognathus* and *Mahengecharax carrolli*, both from the Eocene of Tanzania [MUR 03a, MUR 03b]. *Eocitharinus* was identified as a citharinoidei, a clade that is no longer believed to belong to the characiforms but to form a sister group to the characiforms plus the siluriforms [NAK 01]. Its ancient presence in Africa could be the result of a Cretaceous vicariance [CHE 13] and we can predict their future discovery in the African Late Cretaceous. When it was initially described, *Mahengecharax* was considered to be an alestid, but Malabarba and Malabarba [MAL 10] challenged this identification and prefer to see it as an indeterminate characiform. The African characiforms from the Oligocene and Neogene are relatively common and attributed to the alestid family, currently endemic to the continent [MUR 00, STE 01]. Modern genera have been known since the Eocene, with *Hydrocynus* in Libya [OTE 10] and Egypt [MUR 10], for example.

In South America, Gayet and Meunier [GAY 98] identified isolated remains attributed to six modern taxa of characiforms (families and subfamilies) in the El Molino Formation, which spans the Maastrichtian–Danian boundary. Arratia and Cione [ARR 96] and Malabarba and Malabarba [MAL 10] prefer to consider the fossils to be indeterminate characiforms, because of the extreme diversity of modern characiforms. Another mention of Cretaceous South American characiforms is a ctenoluciid from the Maastrichtian of the Bauru Group in Brazil [MAL 10] that was initially identified as a characiform with affinities to the alestids [GAY 89]. The South American Cenozoic fossil record of characiforms is relatively diversified [LUN 98].

Apart from their distribution in western Gondwana (Africa and South America), characiforms are also known in Europe in the terminal Cretaceous [OTE 08] and in the Cenozoic [GAU 14]. According to Gaudant, three waves of African characiforms succeeded each other during the Cenozoic, with the first during the Ypresian, the second during the Rupelian (*Eurocharax tourainei*) and the third during the Middle Miocene. With the exception of the first migration, which reached the Paris basin, Belgium and southern England, the dispersals that followed seem limited to southern Europe, probably for climatic reasons. It should be noted that the vast majority of European occurrences are isolated teeth. Despite the very characteristic morphology of characiform teeth, the degree of identification is not as high as for mammals. It is not impossible that new discoveries could challenge the scenario of the three dispersals suggested by Gaudant. Finally, there is the presence of isolated dentaries in the Campanian of Alberta, Canada, attributed to a characiform [NEW 09]. This unexpected presence of these fishes is explained by Newbrey *et al.* as a dispersal from Europe or South America in the Late Cretaceous during a hot period. It could also be proposed, based on the respective ages of the Cretaceous characiforms, that the North American form came from South America and illustrates an episode of the dispersal of the alestid plus hepsetid clade (according to Chen *et al.* [CHE 13]) toward Africa. Passages of continental vertebrates from South America toward North America in the terminal Cretaceous have already been proposed. The presence of characiforms in the terminal Cretaceous in Europe, some of which referred to the alestids by Otero *et al.* [OTE 08], constitutes a second episode in this dispersal. This scenario implies a movement of alestids from Europe toward Africa in the terminal Cretaceous, while the majority of the dispersals from this time were assumed to go the opposite way (for example, cf. [CAV 16] for the mawsoniid coelacanths).

According to the phylogeny of Chen *et al.* [CHE 13], the two African families outside of citharoidei, the alestids and hepsetids, form a clade. In that case, the broad strokes of the biogeographical history of the characiforms can be easily explained (Figure 4.10). After a vicariance between the citharoidei in Africa from other characiphysi in South America in the Late Jurassic or Early Cretaceous, the latter diversified in South America. A dispersal of the alestid plus hepsetid clade, possibly a unique event in the history of the group, occurred in the Late Cretaceous because of euryhaline forms that spread to North America, Europe and Africa where they diversified. The European tertiary species were either descendants of this dispersal or the result of new dispersals in the Cenozoic.

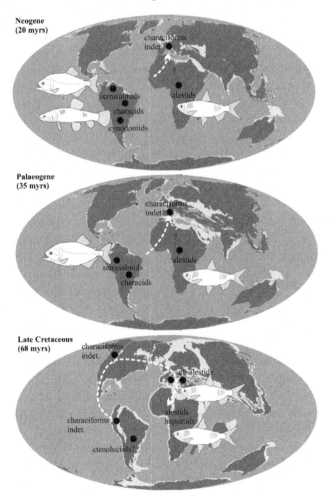

Figure 4.10. *Evolutionary history of characiforms at the end of the Cretaceous and in the Cenozoic. Dispersals are represented by dotted arrows and occurrences by stars. For a color version of the figure, see www.iste.co.uk/cavin/fishes.zip*

4.15.2.2.4. Siluriforms (Siluriformes)

Siluriforms, or catfish, make up the most diversified order of vertebrates [ARM 11]. Today, they are present on all continents except Antarctica and are extremely diversified in South America. The group has several synapomorphies that leave no doubt about its monophyly [FIN 81, FIN 96, ARR 03]. In addition, the presence of dense and thick dermal ossifications as well as articulated solid spines in front of the dorsal fin and pectoral fins in several genera increases the chance of fossilization. These characteristic elements facilitate the identification of their fossil remains, in any case to a high taxonomic classification. The South American continent yielded the greatest number of siluriform fossils, including the oldest known ones. It was very probably that on this continent the order detached from its sister group. An important pending question about the evolutionary history of the siluriforms is to choose between an ancient origin on Pangaea, followed by diversifications through vicariances and dispersals [DIO 04], or a more recent origin followed by dispersals only [GAY 03]. The second solution implies that several siluriform groups previously had the ability to tolerate a marine environment.

Ferraris [FER 07] published a list of siluriforms including the fossil taxa. Not all of the data are addressed here. The fossil record of the siluriforms in the Cretaceous is meager, but it does exist. The oldest known fossils are fragments of spines dating from the Campanian of the Los Alamitos Formation in Argentina [CIO 87]. Initially considered to belong to a diplomystid [CIO 87], Lundberg [LUN 98] challenged this identification by pointing out that the features used are pleisomorphic (diplomystids are in a basal position within the siluriforms). Other siluriforms from the South American Cretaceous were discovered in the Maastrichtian of the Coli Toro [CIO 80] and Yacoraite formations [CIO 85] in Argentina, as well as in the Marília and Adamantina formations in Brazil [GAY 89]. The other South American occurrences came from layers that are difficult to date around the Maastrichtian–Paleocene boundary. Gayet [GAY 88b, GAY 90, GAY 91] and Gayet and Meunier [GAY 98] described three genera, *Andinichthys*, *Hoffstetterichthys* and *Incaichthys*, which are considered to be siluriforms *incertae sedis* by Lundberg [LUN 98]. Cione and Prasad [CIO 02] described the base of a pectoral spine from the Maastrichtian of the Intertrappean Beds in India connected to an indeterminate siluriform. The only mention of siluriforms in the African Cretaceous comes from the Coniacian-Santonian layers in In Becetem in Niger by Patterson [PAT 93] recorded in the "Fossil

Record" Book. Without a more precise description of this fossil, this occurrence is considered provisional [GAY 99]. The fossil record of the siluriforms explodes on a large geographical scale in the Eocene. In the Cenozoic, they are distributed on all continents, including Antarctica. Antarctica yielded the pectoral spine of an indeterminate siluriform dating from the Late Eocene or the Early Oligocene of the Meseta Formation on Seymour Island [GRA 86]. Lundberg [LUN 98] revised the fossil record of siluriforms for the South American Cenozoic while Gayet and Otero [GAY 99] and Gayet and Meunier [GAY 03] synthesized all of the paleontological data on a global scale. Detailed information is not given here.

Based on the fossil record of the siluriforms, it seems clear that their evolutionary radiation occurred during the Paleogene (Figure 4.11). In the Cretaceous, this radiation, or rather these continental radiations, occurred soon after the dispersal of the large clades that made up the order in the Cretaceous, but they left few visible traces in the fossil record, at least at that time. According to Gayet and Otero [GAY 99], the Cenozoic radiation of the siluriforms as a whole is linked to the appearance of key innovations. These innovations, which are probably connected to one another, include the ankylosis of the vertebrae in the Weberian apparatus, a significant mineralization of the skeleton in the anterior part of the body, and the presence of strong spines in front of the pectoral and especially dorsal fins. These apomorphies affected portions of the skeleton that are easily fossilized and identifiable. If taxa with these features are not commonly found in the continental Late Cretaceous deposits, it means that they were absent or very rare. It is possible that stem siluriforms that do not have these apomorphies existed in the Late Cretaceous, but they have not yet been recognized. However, it is not very likely that siluriforms from the crown group (which therefore possessed these key innovations) were present and diversified at this time.

The biogeographical conclusions reached by Sullivan *et al.* [SUL 06] on the basis of their molecular phylogeny are in agreement with fossil data. The South American origin of the order is supported by the fact that the loricarioidei, which are exclusively South American, are situated in a more basal position than the diplomystids (also South American) in this phylogeny. The presence of extinct genera similar to diplomystids in South America, including *Bachmannia* in the Paleogene and *Kooiichthys* in the Neogene, underscore the cradle role of this continent for this order [AZP 11, AZP 15]. The presence of *Hypsidoris*, a genus from the Eocene of North

America that may be close to those two South American genera [GRA 98] could be explained by a dispersal from South America toward North America during the Late Cretaceous and the Paleocene, a dispersal that included various groups of continental vertebrates [GAY 92]. The phylogeny by Sullivan *et al.* [SUL 06] does not present any case of sister clades between South America and Africa. The topology shows an absence of vicariance events but reinforces the hypothesis of dispersals after the opening of the South Atlantic. However, the clear result of this phylogeny is the presence of endemic radiations on three continents: the loricarioidei in South America, a "big Asia" radiation and a "big Africa" radiation that include clades that have not yet been named.

The ictalurids are an endemic North American family that forms a clade with the cranoglanidids, a monogeneric family endemic to China and Vietnam, within the large polytomy of siluroids [SUL 06]. The oldest fossils of ictalurids come from the Paleocene [PAT 93] and are relatively plentiful in the Eocene of the Bridger and Green River formations with the genus *Astephus*, placed in a basal position in relation to other ictalurids [GRA 88]. Detailed information about the fossil record of ictalurids can be found in Smith for the Neogene [NEO 81], Ferraris [FER 07] and Ross [ROS 13]. Ross pointed out that the family remained confined to North America during the Cenozoic despite the presence of connections with Europe and Asia. However, the presence the sister family of cranoglanidids in Asia indicates a passage of representatives of this clade through Beringia at the end of the Mesozoic or the start of the Cenozoic.

One question remains in this scenario. What was the timeline and the route taken during the dispersal of siluroids from South America toward Eurasia and Africa at the end of the Cretaceous or the start of the Cenozoic? The dispersals must have occurred relatively quickly because families of this clade were quickly identified at the start of the Cenozoic on different continents. In Africa, the bagrids are known since the Eocene with *Eomacronies* and *Nigerium* [WHI 36, LON 10] and the claroteids with *Chrysichthys* [MUR 03]. However, the latter generic attribution was challenged by Otero *et al.* [OTE 15], although they recognize the presence of *Clarotes* or a related genus in the Eocene of Libya. The clariids are known in Africa since the Oligocene [GAY 99]. Certain results obtained from the molecular studies are more difficult to integrate in the framework proposed here, such as the origin and diversification of the pimelodoidea in South America in the "mid" Cretaceous [SUL 13]. However, alternative dates, more in accordance with the fossil data, have also been proposed [SAN 09]. The biogeographical history of the loricariids in the Cenozoic is detailed by Roxo *et al.* [ROX 14]. The authors highlight the importance of river captures that allowed fishes to increase their distribution area and, concurrently, show the importance of marine transgressions that contributed to building river networks.

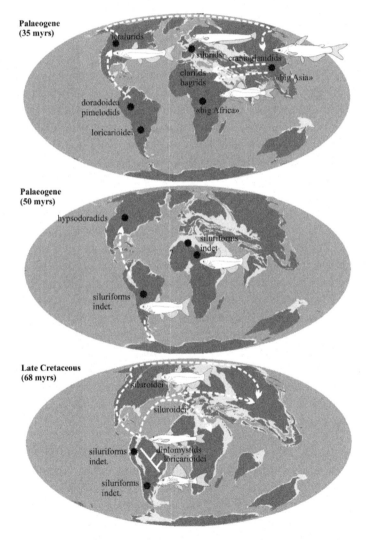

Figure 4.11. *Evolutionary history of the siluriforms at the end of the Cretaceous and in the Cenozoic. Dispersals are represented by dotted arrows and occurrences by stars. For a color version of the figure, see www.iste.co.uk/cavin/fishes.zip*

4.16. Euteleosteans (Euteleosteomorpha)

4.16.1. *Protacanthopterygians (Protacanthopterygii)*

The contents of this taxon have varied since its creation. Currently, it includes four orders. Three of the orders (galaxiiforms, salmoniforms and osmeriforms) occur primarily or partially in freshwater. The fourth order, the argentiniforms, is marine.

4.16.1.1. *Galaxiiforms (Galaxiiformes)*

The galaxiiforms and the salmoniforms have interesting amphitropical distributions. The galaxiiforms are limited to the southern hemisphere and the salmoniforms to the northern hemisphere. The galaxiiforms are restricted to New Zealand and the southernmost parts of Australia, South America and South Africa. Four models have been proposed to explain this distribution: successive dispersals through marine environments, as certain species are tolerant to seawater [MYE 49, MCD 02]; dispersals in continental environments through land bridges ([GIL 93], cited by [ROS 74]); dispersals across Gondwana before its fragmentation; vicariant events in connection with the breakup of Gondwana (both latter hypotheses discussed by Rosen [ROS 74]). Fossils of galaxiiforms are rare and do not provide much information on the evolutionary history of the group, especially its biogeography. Anderson [AND 98] described a galaxiid in the terminal Cretaceous (Maastrichtian) of South Africa, an identification that was challenged by McDowall [MCD 02]. McDowall described galaxiids similar to modern forms in the Miocene of New Zealand [MCD 97, LEE 07]. The schema used here is that of McDowall [MCD 02], taken up by Burridge *et al.* [BUR 12] based on a molecular phylogeny. It considers the modern distribution of galaxiids to be the result of marine dispersals, especially for diadromous forms, although a Gondwanan "deep layer" in the distribution of certain taxa should not be excluded (Figure 4.6).

Lepidogalaxias salamandroides is a small, very particular fish about 7 cm long. It is endemic to a small area in southwestern Australia where it occupies shallow ponds that are very acidic. It is distinguished by its ability to move its head independent of its body because of a flexible "neck" and the ability to tolerate periods of drought by taking refuge inside a burrow dug in the sediment. Long connected to the galaxiids or the osmeriforms, recent molecular phylogenies [LI 10, BET 13] situate it as a sister group to all other euteleosteans (or euteleosteomorpha according to Betancur *et al.* [BET 13]). The phylogenetic position of *Lepidogalaxias*, and also the galaxiids, makes it likely that its lineage already existed in the Early Cretaceous.

4.16.1.2. Salmoniforms (Salmoniformes)

We use the results of Campbell *et al.* [CAM 13], who found a clade grouping the salmoniforms and esociforms, which they suggested naming the Salmoniformes. The "umbrids" do not appear to be monophyletic and are included in the esocid family.

The salmoniforms form a mirror group of the galaxiiforms with a distribution limited to the northern hemisphere (Figure 4.6). The salmoniforms share few unique derived features visible on their skeleton. Several of the supposed "salmoniforms" described in the Cretaceous such as *Barcarenichthys* [GAY 88a], *Spaniodon* [TAV 03] or *Kermichthys* [KHA 10], are probably basal euteleosteans but are not necessarily salmoniforms *sensu stricto*. Similarly, the orthogonikleithrids, small marine fishes from the Late Jurassic, are connected to the "salmoniforms" by Arratia [ARR 97], but this order is understood by the author in a broad sense. However, these occurrences illustrate that the first euteleosteans and probably the first protacantopterygians also were marine.

The oldest esocid fossils as well as the oldest fossils of their sister group, the salmonids, are known in North America in the Late Cretaceous and the Eocene, respectively. We can suggest, cautiously because the fossil record is still sparse, that a common ancestor lived in freshwater on this continent in the Early Cretaceous as suggested by the phylogeny of Campbell *et al.* [CAM 13]. The reconstruction of ancestral habitats from a molecular phylogeny [ALE 13] also indicates a freshwater origin for the common ancestor of the esociforms and the salmoniforms. This study suggests that anadromy appeared several times within distinct freshwater lineages of salmoniforms.

4.16.1.2.1. Esocids

The esocids include the pikes (esocids) and the "umbrids". These two groups are not very diversified in terms of taxonomic variety, but they are widely distributed in freshwater environments in the northern hemisphere.

The genus *Estesesox* is found in the Campanian of Alberta, Canada in the Milk River, Foremost and Oldman formations, with the latter also containing the genus *Oldmanesox*. *Estesesox* can also be found in the Maastrichtian of the stateside formations of Lance and Hell Creek in Wyoming and Montana respectively [WIL 92]. After the Cretaceous–Paleocene boundary, the genus *Esox* was reported in the Thanetian of the Paskapoo Formation in Alberta [WIL 92]. If this occurrence is confirmed to belong to *Esox*, the duration of this genus is noteworthy, at more than 55 million years [ROS 13]. The occurrences of esociforms in the Neogene of North America were summarized in Smith [SMI 81]. In Europe, we find the genus *Esox* in the Early Oligocene of the Apt region of France [GAU 78a], and then in the Late Oligocene of Germany and the Czech Republic, and then in various sites from the Neogene, notably in Öhningen in Germany.

The "umbrids" were traditionally regarded as a family. According to the analysis of Campbell *et al.* [CAM 13], however, the three genera and seven species distributed in Eurasia and North America are placed in basal positions in relation to the esocids and must therefore be grouped with the esocids in a single family. One species, *Dallia pectoralis*, is present on both sides of the Bering Strait. The fossil record of the "umbrids" in Europe begins in the Paleocene with the genus *Boltyshia*, followed by *Palaeoesox* in the Eocene and the genus *Umbra* in the Late Oligocene [GAU 12]. In North America, the genera *Dallia* and *Umbra* are known in the Neogene [SMI 81]. According to the calibrated phylogeny of Campbell *et al.* [CAM 13], the lineage of *Umbra*, the most basal genus of esocid, would have separated from the other members of the family, including *Esox*, in the Late Cretaceous 88 million years ago. This schema is in accordance with the fossil record of the *Esox* lineage, but signifies a lack of paleontological information for the "umbrids".

4.16.1.2.2. Salmonids

The most ancient salmonid *sensu stricto* is *Eosalmo driftwoodensis* described by Wilson [WIL 77]. It is known in various sites from the Middle Eocene of British Columbia in Canada and Washington, USA [WIL 99]. These authors showed that *Eosalmo* was already deeply nested within the salmonids. It represents the sister group of all salmonins. The occurrences of salmoniforms in the Neogene of North America were summarized by Smith

[SMI 81]. Ross [ROS 13] suggests that the modern primarily Eurasian genus *Salmo* and the genus *Onchorhynchus* from the west coast of North America resulted from a vicariance phenomenon in Beringia. However, this hypothesis has been jeopardized by recent phylogenies like the one by Macqueen and Johnston [MAC 14], which shows that the genus *Salvinius*, present in the Palearctic and Nearctic domains, is the sister genus of *Onchorhynchus*. Various formations from the Late Miocene to the Pleistocene of the states of Idaho and southeastern Oregon yielded a wide variety of salmonids alongside ictalurids, several cyprinids, catostomids, centrarchids and cottids [STE 16]. The diversification of the salmonids is associated with the appearance of anadromy and is linked to a drop in temperature in the Neogene [ALE 13, MAC 14]. The increase in the diversity of the family over time follows an exponential curve comparable to the increase in diversity of freshwater fishes as a whole, and not a logistical mode of increase as observed in marine fishes (cf. 5.2.1).

4.16.1.3. Osmeriforms

The osmeriforms include four families: smelt (osmerids) in the northern hemisphere, plecoglossids in Asia, retropinnids in the Australian region and salangids in southeastern Asia. All of these families are euryhaline and include some freshwater and diadromous species. The oldest known osmerid fossil is *Speirsaenigma lindoei* in the Paleocene of the Paskapoo Formation in Alberta, Canada [WIL 91]. These authors found that the phylogenetic position of this species is relatively derived within that family, that is a sister group of *Plecoglossus*. This diadromous genus from the northwest of the Pacific Ocean has a few populations blocked in continental environments. In Europe, the osmerids are known from the Early Oligocene of Alsace and Provence with the genus *Enoplophthalmus* [GAU 13], as well as from the Oligocene of Rott in Germany. They indicate, according to Gaudant [GAU 85], a substantial drop in temperature in the climate of the Oligocene.

4.16.2. Paracanthopterygians (Paracanthomorphacea)

4.16.2.1. Percopsiforms (Percopsiformes)

Today, this order includes the amblyopsids, or cavefishes, which comprise some cavernicolous species with vestigial eyes, as well as aphredoderids, which have a single species, and percopsids, which also have only one species [DIL 11]. The extant species are endemic to North America, as are the seven fossil genera connected to them (Figure 4.6). Murray and Wilson [MUR 99] proposed a phylogeny of the group that includes both fossil and living forms. According to this study, the amblyopsids are not directly connected to the percopsiforms. All the fossil taxa, however, are included within the percopsiforms. The most basal form, which is also the most ancient (with the exception of isolated fragments attributed to this order found in the Hell Creek Formation from the terminal Cretaceous by Brinkman *et al.* [BRI 14]), is *Mcconichthys longipinnis* from the Early Paleocene of the Tullock Formation in Montana. Next came *Libonichthys* from the Middle Eocene, with *L. blakeburnensis* in British Columbia and *L. pearsoni* in Washington, which form the sister group of the two modern families. The sister group to *Aphredoderus* is *Trichophanes*, in the Late Eocene of the Florissant Formation in Colorado. Within the percopsids, *Massamorichthys* is from the Late Paleocene of the Paskapoo Formation in Alberta and is the sister genus to the modern *Percopsis*. Two genera from the Eocene of Wyoming, *Amphiplaga* and *Erismatopterus*, form a sister clade to the previous one. *Lateopisciculus*, from the Paskapoo Formation, is situated in trichotomy with the two pairs of genera. Recently, an indeterminate amblyopsid was found in the Late Eocene–Early Oligocene of the Cypress Hills Formation in Saskatchewan [DIV 15]. This mention was followed by a second one from the Wasatch Formation in Wyoming dating from the Early Eocene [DIV 16]. The interpretation of this material, made up of isolated elements, suggests that the two percopsids from Green River, *Amphiplaga* and *Erismatopterus*, could in fact be amblyopsids. The paleobiogeographical signal of this fossil record is small on a global scale because it deals with intracontinental relations that are not discussed here. The fossils illustrate, however, the existence of an order endemic to the North American continent throughout the Cenozoic, which seems to have had a small radiation in the Paleogene.

4.16.3. *Percomorphs (Percomorphacea)*

4.16.3.1. *Gobiiforms (Gobiiformes)*

Gobies are a highly diversified family (about 2,000 species) that is primarily marine, although it also contains species that live in brackish environments and a few freshwater species. In the Oligocene of Europe, for example, *"Gobius" aries* is found in the Marseille [GAU 75] and Aix-en-Provence basins [GAU 78b]. Following the discovery of well-preserved fossils in Aix-en-Provence, the species was placed in the extinct genus *Lepidocottus* within the butid family [GIE 13]. This family is currently absent from the Mediterranean region. McDowall *et al.* [MCD 06] noted the presence of the modern eleotrid genus *Gobiomorphus* in the Early Miocene and the Pleistocene of New Zealand. They point out that this genus does not show a significant diversification despite its ancient existence, contrary to the galaxiids that were also present in this biogeographical province but which are very diversified. The freshwater species of gobiiforms are probably distinct "captures" in different regions of the world and so do not constitute a large-scale biogeographical indicator.

4.16.3.2. Anabantaria (Anabantaria)

4.16.3.2.1. Channids (Channidae)

The channids are a family that currently includes two genera, one present in Africa (*Parachanna*) and the other in Asia (*Channa*). They are notably distinguished by the presence of a suprabranchial organ that allows them to breathe air. The fossil record of the family begins at about the same time in Asia and Africa. In Asia, it consists of *Anchichanna* [MUR 08] and *Eochanna* [ROE 91] from the Eocene of Pakistan (the two taxa may be cospecific but it is not possible to prove it for the moment because the preserved anatomical parts do not overlap). In Africa, it consists of *Parachanna* in the Eocene of Libya [OTE 15] and the Eocene and Oligocene of Egypt [MUR 06]. A slightly younger genus, *Parachannichthys*, was described in the Indian Miocene [GAY 88c]. Li *et al.* [LI 06] suggested, based on a molecular analysis, a divergence between the two modern genera in the Early Cretaceous, then a dispersal from Gondwana to Asia on the Indian "raft", as has also been proposed for two clades of osteoglossiforms. However, the fossil record and new estimates for the age of the divergence [ADA 10] contradict this hypothesis and instead suggest that a dispersal took place, probably in the Paleogene (Figure 4.6). Madeleine Böhme [MAD 04] discussed the distribution of channids and presented an interesting model connected to the climate. She observed that between 17 and 13 million years ago, channids were present, in addition to their supposed origin point in Pakistan, western Europe (north of the Pyrenees and north of the Alps), Asia and subtropical Siberia. After the start of the Miocene, between 13 and 8 million years ago, channids are no longer found in Europe, but were exclusively located in southern and central Asia. Then, since the Messinian 7 million years ago, they are found in their current distribution area, which is eastern Asia and Africa. These variations in distribution are associated with the variations of a subtropical climate linked to strong summerlike precipitation, in particular to the start of the Asian monsoon, rather than to physical opportunities for continental passage. The preponderance of climatic features on the only possible physical paleogeographical connections observed in the evolutionary history of channids is also found in the history of other groups of freshwater fishes.

4.16.3.3. *Carangaria (Carangaria)*

4.16.3.3.1. Latids (Latidae)

This family, as it was defined by Otero [OLG 04], contains two modern genera (*Lates* and *Psammoperca*), as well as the extinct polyphyletic genus *Eolates* that includes the stem species of the family. According to Otero [OTE 04], *Eolates* includes three European species. Two are marine and the other, *E. aquensis*, is lacustrine. It is known in the Late Oligocene of Aix-en-Provence. The genus *Lates* has a rich fossil record in Africa and in Europe summarized by Otero [OTE 04]. It is recorded from the Eocene of Birket Qarun in Libya by Arambourg in 1961 [OTE 15]. The fact that the most basal and oldest species of the family, *Eolates gracilis*, as well as some species of the genus *Lates* (*Lates calcarifer*), but also the sister genus of *Lates*, *Psammoperca*, all occupy marine coastal environments is a strong indication of dispersals in marine or coastal environments at the beginning of the Cenozoic. Following these dispersals, several species were then blocked into continental environments.

4.16.3.3.2. Ambassids (Ambassidae)

The ambassids include several euryhaline genera currently present in Asia, Oceania and the West Pacific-Indian Ocean. This family is present in the European Cenozoic in deposits of freshwater or brackish origin. *Dapalis* is relatively common in the Late Eocene and the European Oligocene. The genus was first connected to the centropomid family [GAU 77, GAU 81], then to the chandids [GAU 87], a family that is now synonymous with the ambassids. The ambassids are still present in the Early Miocene of the western Paratethys [BÖH 03].

4.16.3.3.3. Cichlids (Cichlidae)

Cichlids hold a particularly important place in the faunas of freshwater fishes for two reasons. On the one hand, their extraordinary diversification in the lakes of the East African Rift makes them an exemplary case study for understanding the mechanisms at the origin of evolutionary radiations. On the other hand, their current geographical distribution is generally considered to be a classic example of vicariances related to the fragmentation of Pangaea. The family is found on continental masses of Gondwanan origin, such as Africa, the Middle East, Madagascar, India and South America, but it is absent from Australia and Antarctica. It is also found in non-Gondwanan regions that have only emerged since the Neogene such as Iran, the southern part of North America and the Antilles. The fossil record of the cichlids was revised by Murray [MUR 01b] and is summarized here. The most ancient representatives are *Mahengechromis* from the Middle Eocene of Tanzania [MUR 00] and a form similar to *Tylochromis* in the Middle Eocene of Libya [OTE 15]. Then, fossils are known in the Oligocene of Africa, in Somalia and Saudi Arabia. In the Miocene, the distribution area of the cichlids increased with occurrences in Uganda, Kenya, Tanzania, Algeria and Tunisia, as well as in South America and Europe (Italy, Germany and Switzerland). The presence of cichlids in Europe during the Miocene resulted from freshwater ichthyofauna exchanges in the Middle Miocene. These exchanges concerned, in addition to the cichlids, the clariids, characids and latids [OTE 01b]. One of these European species from the Late Miocene of Italy, *Oreochromis lorenzoi*, is close to the modern species *Oreochromis mossambicus* [CAR 03]. It is likely that the extinct species had the same ability as the current species to tolerate brackish water, which allowed it to move from Africa to Europe through coastal and lagoonal environments. In the Pliocene, remains are known in various countries in Africa and South American, as well as in Haiti. Murray [MUR 01b] pointed out that the fossils of cichlids are relatively common in the continental layers since the Eocene but are completely absent in older layers. She concludes that the group probably did not exist in the Cretaceous. She proposes, based on an interpretation of a phylogeny, an origin in Madagascar during the Paleocene followed by dispersals toward Africa and toward India. From

Africa, a first dispersal occurred toward South America (across the Atlantic Ocean, which would have been about 500–800 km wide) and a second toward Iran. Since Murray's work, new cichlid fossils have been described in the Eocene of the Lumbrera Formation in Argentina, *Proterocara*, *Gymnogeophagus* and *Plesioheros* ([MAL 06, MAL10] and [PER 10], respectively). The debate about the respective roles of vicariance and dispersal in the current distribution of the cichlids, which began several decades ago, continued after Murray's publication. Briggs [BRI 03], for example, proposed dispersals across rather large inlets by justifying that these fishes, or at least some of them, can tolerate seawater. Sparks and Smith [SPA 04] vigorously criticized Briggs' perspective, mostly on a methodological basis. They reproached Briggs for proposing a narrative hypothesis that, consequently, cannot be disproved. This criticism is recurrent for biogeographers who draw their conclusions exclusively from the interpretation of phylogenies, without taking the fossil record into account (except to calibrate the molecular clocks). They only retain vicariance as the explanatory model at the origin of biogeographical distributions. The criticism of "indemonstrable narratives", when alluding to dispersals, is a critique that is based on the fact that the absence of a fossil of a clade does not necessarily signify the actual absence of this clade. This question was approached in a constructive way in a recent article about cichlids. Friedman *et al.* [FRI 13] tested "the orthodoxy of the Mesozoic vicariance" by considering that the absence of paleontological proof could be considered to be evidence of an absence in statistical terms. They analyzed the distribution of the fossiliferous areas containing cichlids in combination with an empirical function that quantified the potential to discover fossils in an interval that is credible for the appearance of a clade. For cichlids, this interval is situated in the Paleocene, between 65 and 57 million years ago, and therefore far from the date that involved a distribution linked to the fragmentation of Pangaea in the Early Cretaceous. Therefore, it appears more probable, in relation to the presence of cichlid fossils in the geological layers as well as their absence in older layers, that dispersals would be at the origin of the current distribution of the family (Figure 4.12). The model of the dispersals retained here is that of Murray [MUR 01b], a model that was recently mostly confirmed by Matschiner *et al.* [MAT 16]. In that study, the calibration of the phylogeny was carried out by defining the calibration densities based on the first fossil occurrences to which was added an estimate of its sampling rate.

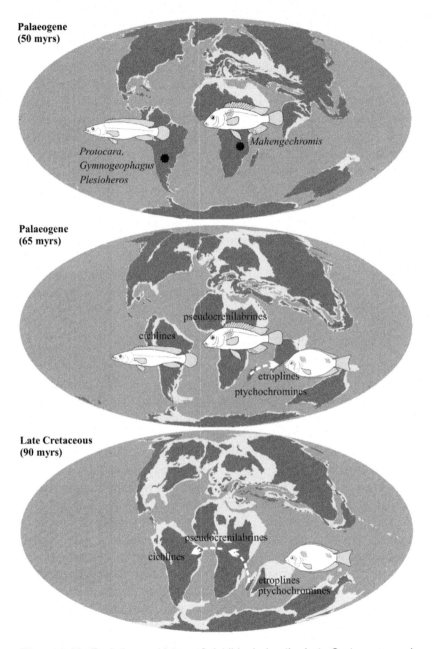

Figure 4.12. *Evolutionary history of cichlids during the Late Cretaceous and Paleogene. Dispersals are represented by dotted arrows and occurrences by stars. For a color version of the figure, see www.iste.co.uk/cavin/fishes.zip*

4.16.3.3.4. Atheriniforms (Atheriniformes)

This order mainly contains marine species, but some are freshwater, including *Chirostoma*, an atherinopsid that has a species flock in some Mexican lakes [BAR 73]. *Craterocephalus*, an atherinid from Australia, New Guinea and East Timor is also very diversified in freshwater [UNM 10]. It represents a typical case of freshwater occupation by a primarily marine family where the primary freshwater fishes are not diversified. This role as a "vicarious" species was probably already played by atheriniforms in the Paleogene of Europe where the genus *Palaeoatherina* is known in the Late Eocene and the Early Oligocene [GAU 89]. The genus *Hemitrichas*, initially described based on otoliths, is known from the Eocene to the Miocene of Europe in environments with very variable salinity [GAU 05]. Recently, a molecular phylogeny of a large sample of species of atheriniforms, accompanied by a reconstruction of ancestral environments, clearly indicated that the distribution of the species was due to marine dispersals with several cases of capture in continental environments [CAM 15].

4.16.3.3.5. Cyprinodontiforms (Cyprinodontiformes)

These small fishes, some of which are ovoviviparous or viviparous, are found in environments where the water quality can vary greatly (salinity, temperature or acidity). Certain genera form species flocks. The aplocheiloid clade has a tropical distribution with the aplocheilid family in Asia and Madagascar, nothobranchiids in Africa and rivulids in South America. A phylogeny proposed by Murphy and Collier [MUR 97] found the African and South American families to be sister groups, and the Malagasy and southern Asian clades in a sister position to the western Gondwanan clade. This topology reflects a double vicariance linked to the fragmentation of Pangaea. However, a more recent analysis [POH 15] found a sister relationship between the Asian and African families, with the South American family being situated in a sister position to this pair. This new topology is more in accordance with the position of the cyprinodontiforms nested within the percomorphs, a position that contradicts the ancient vicariance phenomena linked to the first phases of the fragmentation of Pangaea. It suggests instead that dispersals occurred between the tropical zones, and in particular between Africa and Asia during the Cenozoic at the same time as other freshwater fishes (Figure 4.6). The only aplocheiloid fossil is *Kenyaichthys kipkechi* from the Late Miocene of Kenya [ALT 15]. This species was placed in its own family and provides little information about the biogeographical history of the suborder. The sister group of the aplocheiloids is that of the cyprinodontoidei distributed primarily in North Africa, Eurasia and Africa. The ability of several species of cyprinodontiforms to tolerate brackish environments explains this fishes' capacity for intercontinental dispersal. Hrbek and Meyer [HRB 03] propose an interesting paleobiogeographical scenario to explain the distribution of modern species in the near-eastern and circum-Mediterranean genus *Aphanius*. The speciation within this genus is the result of the closing of the Tethys around the Oligocene–Miocene boundary. They note the agreement between the spatial distributions and the phylogeny, indicating vicariance phenomena, but they also point out the importance of environmental characteristics for understanding the modern distribution of the species. The fossil record of the European lineage of the cyprinodontiforms is abundant. It begins in the Oligocene with the common genus *Prolebias* present in the Aix-en-Provence basin [GAU 81a], the Haute-Provence Alps [GAU 78] and Limagne [GAU 12]; in Germany in Kleinkems [GAU 81], in Serreal in Spain [GAU 82]; and in saliferous zones from the Early Oligocene of Alsace. This genus is typical of freshwater or slightly brackish environments, which allowed Gaudant [GAU 81] to reconstruct a slightly saline environment for certain layers of the saliferous zones in Alsace. Costa [COS 12] revised the phylogeny of the order by including six extinct taxa with various living species. This analysis led him to recognize the original genus *Prolebias* represented by a single species within the valenciid family, and two new genera, the valenciid *Francolebias* and *Eurolebias*, which he placed in the cyprinodontid family. He also describes two new extinct species of the modern genus *Pantanodon*, a poeciliid (but see [POH 15] for the phylogenetic position of this genus represented by two species from East Africa and Madagascar.) *Francolebias* presents a sexual dimorphism at its anal fin, which indicates the presence of a copulatory organ in males and, consequently, an ovoviviparous or viviparous mode of reproduction. A new species of *Francolebias* was recently recognized in the Early Oligocene of Chadrat, in Puy-de-Dôme in France [GAU 16]. These data indicate that the diversity of the cyprinodontiforms in Europe was high during the Oligocene and included lineages currently limited to the American and African tropical zones. The fossil record of the cyprinodontiforms is quite diversified in the Neogene of North America, South America and Europe, but it is not detailed here. We will just mention the presence of *Carrionellus diumortuus* in the Early Miocene of Ecuador that belongs to a genus that is very similar to

Orestias, a modern cyprinodontiform that lives in high-altitude lakes in the central Andes where they form a species flock [COS 11].

4.16.3.4. *Eupercaria (Eupercaria)*

4.16.3.4.1. Moronids (Moronidae)

This family includes two modern genera that live in freshwater, brackish and marine environments in the North Atlantic and coastal rivers. We know of fossils as early as the Middle Eocene of Messel in Germany (*Palaeoperca* and *Rhenanoperca*), the Late Eocene or the Early Oligocene of Cypress Hills in Canada, the Oligocene of Provence and then in various sites in the Neogene of Europe and North America [SMI 81]. The brackish and coastal lifestyle of these fishes does not allow us to ascertain a biogeographical scenario.

4.16.3.4.2. Centrarchiforms (Centrarchiformes)

The contents of this clade, as it is retained here, do not correspond to the classification by Betancur *et al.* [BET 14] but rather follow the results of a study by Chen *et al.* [CHE 14] that was also based on a molecular phylogeny. According to this work, the percichthyids are a sister group to a clade containing the centrarchids, elassmomatids and the sinipercids. In this topology, the two clades are antitropical, with the percichthyids in the southern hemisphere (South American and Australia) and the second clade in the northern hemisphere (centrarchids and elassmomatids in North America, sinipercids in East Asia). This calibrated phylogeny suggests an origin for the clade around the Cretaceous–Paleogene boundary and implies a single transfer from a marine environment to freshwater. However, as the authors recognize, this distribution does not agree with the tectonic development during the Cenozoic and they do not exclude the possibility that the basal forms were euryhaline.

The centrarchids are a freshwater family endemic to North America, known as sunfishes, which includes eight genera and 38 species today. The fossil record is also limited to this continent. It dates back to the Eocene with isolated elements attributed to this family from the Wasatch Formation in Wyoming [DIV 16] and the Cypress Hills Formation in Saskatchewan. The occurrences of centrarchids in the North American Neogene are summarized in [SMI 81]. According to a calibrated molecular phylogeny [NEA 03], the first dichotomies at the origin of the modern genera date back to the Oligocene.

To my knowledge, the most ancient percichthyid is *Percichthys hondoensis* from the Maastrichtian of the El Molino Formation in Bolivia [GAY 98]. Next, *Percichthys hondoensis* is found in the Eocene of the Cañadón Hondo Formation in Argentina [ARR 96] and *Macquaria antiquus* in the Eocene of the Redbank Plains Formation in Australia [UNM 01]. In Europe, some "percichthyids" were reported in the Oligocene with *Properca* and *Dapaloides* in France [GAU 79b, GAU 89, GAU 84] and in the Eocene of the Duero Basin in Spain [GAU 84b]. However, the definition of percichthyids used by Gaudant does not correspond to the modern definition of this family and these occurrences cannot be taken into account in the evolutionary history of the clade. The record of this family is quite abundant in the Neogene of China and Japan (summarized by Yabumoto and Uyeno [YAB 00]) where it is exclusively of continental origin, with the exception of *Inahaperca taniurai* discovered in the shallow, marine Middle Miocene of the Iwami Formation in Japan. The fossil record agrees with the scenario proposed by Chen *et al.* [CHE 14]: for the southern hemisphere, it was either dispersals through Antarctica or a vicariance event at the origin of the distribution of the percichthyids in South America and Australia; for the northern hemisphere, a dispersal through Beringia in the Paleogene at the origin of the distribution of the centrarchids, elassmomatids and sinipercids.

4.16.3.4.3. Perciforms

The suborder of the cottioids is primarily marine but some species reached freshwater in Asia and North America. A species flock is present in Lake Baikal today. The two North American genera, *Cottus* and *Myoxocephalus*, seem to have reached freshwater independently [ROS 13]. The occurrences of cottids in the Neogene of North America can be found in [SMI 81].

Evolutionary Patterns in Freshwater Fishes

5.1. Vicariances and dispersals

The process of vicariance is the reasonable starting hypothesis for all biogeographical analyses explaining a disjointed distribution because it involves the least amount of assumptions. However, when studying a biogeographical scenario, it often appears that the initial hypothesis is not supported by the data, whether the data come from phylogenetic analyses or paleontological occurrences. Therefore, as we observed previously, including dispersal events is necessary for explaining several distribution patterns, even for ancient clades like the osteoglossomorphs. Including dispersal events when building paleobiogeographical scenarios is often considered to be a methodological weakness because these events cannot be directly observed in the fossil record. For primary freshwater fishes, dispersals across marine barriers involved exceptional phenomena such as displacements on "freshwater rafts", very strong discharges of freshwater by large rivers or other extraordinary events (transport by other organisms, tornadoes, etc.). These rare events are not impossible throughout geological time, but they are indemonstrable and are therefore to be avoided during paleobiogeographical reconstructions. There is also another methodological problem that is often considered unacceptable by biogeographers when constructing scenarios. It consists of considering the absence of a fossil in a given time and place as a usable observation, as was the case in several of the scenarios presented before. These absences are considered to be evidence, or at least indications, of real absences. Although it is true that the statement "the absence of evidence is evidence of absence" is problematic in itself and poses problems when discussing recent periods and very specific locations, it is admissible for much larger scales of time and space (for example both the fossil absence of ostariophysans in the Paleozoic and of cypriniforms in South America correspond to real absences). An innovative quantitative approach that should be applied to a variety of clades is to estimate the probability of the presence of a taxon in fossiliferous layers according to a method used by Friedman *et al.* [FRI 13] for the cichlids (cf. 4.16.3.3.3). This method provides the interval of time that is considered most credible for the appearance of a clade on the basis of the distribution of horizons containing fossils of this clade in a specific area.

Currently, the methods used to reconstruct the biogeographical history of a clade take into account both dispersal and vicariance models, as in the Dispersal-Vicariance analysis method (DIVA). This principle was used in a qualitative way in this work. That said, there are well-documented vicariance events observed in the evolutionary history of freshwater fishes during the last 250 million years of the Earth's history. Most are connected with the first phases of Pangaea's fragmentation. According to our scenarios, they mainly affect the first dichotomies of the large clades of primary freshwater fishes. This includes vicariant events provoked by the division of Laurasia and Gondwana that split the basal osteoglossomorphs from the osteoglossiforms within the osteoglossomorphs (scenario II, Figure 4.7), and split the cypriniforms from the characiphysi within the otophysans (Figure 4.8). The opening of the South Atlantic is associated with vicariant events splitting the citharoidei and the characiphysi (not including the gymnotiforms) within the otophysans (Figure 4.8), and with dichotomies affecting lower ranks of taxonomic groups, such as between genera of chanids, lepisosteiforms, lepidosirenids and mawsoniids (Figure 4.2). One possible intra-Eurasian vicariant event is the split of the cyprinids into the leuciscins and the gobionins, which may have been provoked by the presence of the Turgai Strait (Figure 4.9). The quasi-totality of the other biogeographical patterns can only be explained by dispersals, even for clades whose distribution is traditionally considered to be the result of vicarieant events due to the fragmentation of Pangaea (osteoglossiforms, siluriforms, characiforms, cichlids). The most spectacular examples of dispersal concern the siluroids, alestids, osteoglossins, heterotins and cichlids. These dispersals involved intercontinental crossings through the intermediary of euryhaline forms.

The obstacles facing the dispersals of freshwater fishes are not exclusively linked to the spatial extension of marine barriers. They should also be sought in the biotic and abiotic factors that facilitate or prevent displacements by acting like filters [ROS 13]. One example of this is interspecific competition, which can render an ecological niche inaccessible in a newly attained geographical region because it is already occupied. The physiological characteristics of the organisms concerned also play an important role, such as the capacity to breathe air or move around out of the water. In some cases, however, it is difficult to understand why the biogeographical histories of groups that seem to have similar characteristics are so different. For example, why were the lepidosirenids and the protopterids the passive actors in a vicariant event between South America and Africa during the Cretaceous while at about the same time, the siluriforms were actively dispersing throughout the entire world? After all, lungfishes and catfishes share several characteristics that should make the two groups well adapted to disperse, such as resistance to drought and the abilities to breathe air and move around on solid ground. One possible general explanation is linked to the significant ecological plasticity of the siluriforms compared to lungfishes as demonstrated by several examples of "captures" of catfishes in new geographical areas: *Heteropneustes* in southern Asia [RAT 14], loricariids in South America [ROX 14] and sisorids in the Tibetan plateau [ZHO 16].

An original hypothesis that includes both vicariances and dispersals was proposed to explain the African-Asian and Australian-Asian distribution of some groups. It is based on two major tectonic events that created belated links between Gondwana and Laurasia, namely the detachment of India and then Australia from Gondwana in the Early Cretaceous followed by the collision (India) and closing-up (Australia) of these continental masses with Asia in the Cenozoic. These tectonic events make it possible to propose initial vicariances during the fracturing of Gondwana, then "passive dispersals" while the continental masses moved northward, and finally small active dispersals following contact between India and Eurasia or through Insulinde when Australia got close enough to Sundaland. This general schema makes it possible to explain the continuous distributions between Laurasia and Gondwana by citing vicariant and dispersal events that are mostly passive (the Indian and Australian "rafts"). This scenario has been proposed several times to explain the distribution of osteoglossins [KUM 00], notopterids [INO 09], channids [LI 06] and cichlids [SPA 04]. In these four examples, the raw data is based on calibrated molecular analyses that always suggest extremely ancient ages for these dichotomies compared to the paleontological data. As already observed here, these models, although appealing, cannot withstand factual analysis. The discontinuous distributions of several clades between Africa and Asia can be better explained by dispersals between these two regions during the Paleogene, when the climatic continuity was stable and continental continuity was in the midst of being established. These exchanges occurred in opposite directions depending on the groups (Figure 4.6). Then, this large distribution area was cut off following the aridification of the Arabian plate and a part of central Asia.

Too often, the concept of vicariance relies only on spatial relationships between geographical zones occupied by the organisms. It takes physical surfaces into account without always considering the environmental conditions that characterize them. With freshwater fishes, however, it is likely that the distribution areas were primarily shaped by environmental and climactic forces rather than just being constrained by marine barriers. This situation is ancient and can already be observed during the Triassic on Pangaea, as Schaeffer noted in 1984 [SCH 84] for the redfieldiiforms. The areas of distribution clades identified in the Triassic correspond to surfaces situated along the same general latitude. It should be noted that for the most part, these distributions correspond to the wettest zones on Pangaea, according to a rainfall map (Figure 4.2). Then, during the fragmentation of the supercontinent, the distributions of the clades continue to be situated on the same general latitudes, which, when they break apart, are at the origin of the vicariant events (Figure 4.2).

5.2. Evolutionary radiations

5.2.1. *Causes of the diversification of freshwater fishes*

As mentioned in the introduction, the global diversity of freshwater fishes has increased exponentially since the Late Jurassic while the diversity of marine fishes has followed a logistic curve (Figure 5.1). For freshwater fishes, the growth seems to increase constantly while for marine fishes, after a rapid increase, the curve changes direction and tends to stabilize [GUI 15a]. This stabilization of the diversity in marine environments is particularly distinct for the elasmobranchs (rays and sharks). It can also be observed, albeit less distinctly, for marine actinopterygians and for the group of marine actinopterygians plus elasmobranchs. These two growth patterns of biodiversity correspond to two proposed theoretical models that describe the evolution of biodiversity for the living world as a whole. A logistic curve corresponds to a equilibrium model that considers the Earth system, or a specific ecosystem, to be a closed system that contains a limited biological capacity. This hypothesis arises directly from the theory of island biogeography proposed by MacArthur and Wilson [MAC 63, MAC 67]. An exponential curve, on the other hand, corresponds to a model in continuous expansion and indicates that the observed system does not have a maximum limit to the biodiversity that it can contain, at least for the period under consideration. When compiling the fossil record of invertebrates, Sepkoski [SEP 78, SEP 79] observed three successive diversifications during the Phanerozoic that followed a equilibrium model (logistic). These three episodes were separated from each other by mass extinctions. Benton [BEN 97] performed Sepkoski's exercise again while distinguishing different groups of organisms. He also found a multiple logistic model in marine environments, including marine invertebrates, but in continental environments he observed an exponential increase in biodiversity, particularly in non-marine tetrapods. Observations made about fishes show a similar distinction depending on environment. However, because the analysis compares different lineages within a single group, the actinopterygians, it suggests that the difference observed in diversity dynamics is not only due to the biological or physiological features proper to these taxa, but that it corresponds to distinct evolutionary responses under the control of environmental factors. Or, to be more specifc, it appears that the exponential diversification of freshwater fishes, during the Cenozoic at least, corresponds to the response of a peculiar group, the otophysans, to the environmental pressures that affected it. We can assume a similar situation for other clades in the Mesozoic, but the fossil record is not complete enough to prove it at the moment. As shown by the present-day low diversity of various ancient groups of freshwater actinopterygians, it seems that once a level of equilibrium was attained, the diversity of several clades decreased. Therefore, the exponential shape of the curve observed in continental environments is the result of an accumulation of logistic increases in various clades at various times. At present, this increase is mainly caused by the radiation of otophysans for primary freshwater fishes.

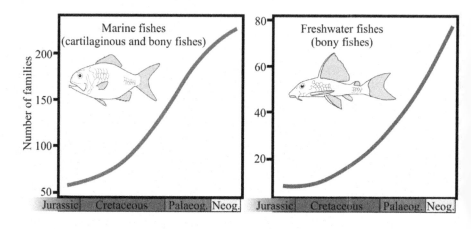

Figure 5.1. *Phylogenetic diversity of marine "fishes" (left) and freshwater actinopterygians (right). The curve on the left is logistic and the curve on the right is exponential (modified from [GUI 15b]). For a color version of the figure, see www.iste.co.uk/cavin/fishes.zip*

What are the possible causes that explain the exponential increase in biodiversity in continental environments? Probably the most obvious one is the fact that the continental environment in general, but especially in freshwater environments, is much more divided spatially than the marine environment. This division fosters the isolation of small populations, which is favorable to speciation events [MAY 94, GRO 12]. The total volume of continental water, much less than the volume of marine water globally, could also be a factor that boosts diversification because in a more limited area, competition between species is a factor of speciation through specialization. Another potential factor is that freshwater species were less affected by the mass extinction at the end of the Cretaceous than marine fishes were [CAV 02, GUI 15b]. This difference is based on the structure of food chains specific to both types of environments: in marine environments, the food chain relies on primary producers while in freshwater environments, the food chain relies largely on detritus (cf 5.3.2). Therefore, in marine environments, the trophic levels occupied by fishes are more sensitive to variations in primary production, in the euphotic zone at least, while freshwater fishes display a greater resilience to these variations. Freshwater fishes would have better overcome the catastrophic event at the Cretaceous–Paleogene boundary. This hypothesis, however, hardly explains the difference in growth direction between the two environments observable in the Cenozoic. Finally, one last hypothesis partly connected to the first hypothesis: during the period considered, which corresponds to the fragmentation of Pangaea, it is possible that the separation of the continental surfaces could have contributed to the multiplication of environments, boosting the diversification of freshwater fishes.

5.2.2. Examples of evolutionary radiations

5.2.2.1. On the scale of continents and large clades

Today, the geographical distribution of the large clades of otophysans is clearly structured on a continental scale: the cypriniforms are present in every part of the world except Antarctica and South America, the siluriforms are present everywhere except Antarctica with a wider diversity in South America, the characiforms are present in South and Central America, Africa, India and Madagascar, and finally, the gymnotiforms are exclusively present in South America. These large diversifications on the scale of entire continents occurred in various freshwater environments. The fossil record informs us of how and when these otophysan diversifications occurred. But are there similar radiations present in extinct clades at a continental scale?

In the Triassic, radiations are present on vast continental areas but they are difficult to detect because the phylogenetic relationships of the groups concerned are generally poorly resolved. In addition to this, the existence of the supercontinent Pangaea, which connected a large portion of emerged lands, makes it difficult to distinguish paleogeographical areas identifiable by clearly defined barriers. However, several groups of freshwater fishes display groupings that indicate the existence of large biogeographical provinces. Redfieldiiforms, for example, present a double radiation on what would become southern Gondwana (South Africa and Australia) and an area covering North America and North Africa. The area of endemism observed in the redfieldiiforms of southern Gondwana is also found for the freshwater perleidiforms, but for them it also includes the southern part of South

America. These fishes are represented in South Africa and Australia with the cleithrolepids, in the southern part of South America with the pseudobeaconiids and, eventually, in South Africa with *Meidiichthys*, the sister taxon of this family. The scanilepiforms of central Asia are among the other radiations in the Triassic covering a large, although still poorly defined, geographical area.

5.2.2.2. On the scale of regions and families (species flocks)

Currently, evolutionary radiations known as species flocks affect very specific clades in relatively limited geographical areas, generally a lake or a small series of lakes. The large body of literature that deals with the modes, rates and mechanisms of recent evolutionary radiations is not addressed here. Today, not all lakes have species flocks, but only about a dozen lakes, all significantly older than the average lake, are concerned [MAR 97]. The evolutionary histories of a few modern clades with species flocks are known.

The cottoids, for example, have a species flock in Lake Baikal in Russia. It concerns pelagic and deep benthic species that evolved from shallow benthic forms [KON 03]. The radiation, for which there is no fossil record, began in the Pliocene or at the start of the Pleistocene according to molecular analyses. The fossil record, however, indicates that a suborder also produced a small species flock in the Pliocene with several species of the genera *Myoxocephalus* and *Kerocottus* found in the Glenns Ferry Formation in Idaho in the United States [SMI 81].

The haplochromine cichlids are the most well-known and most studied cases of species flocks, especially in the East African Great Lakes. The evolutionary radiations, sometimes sympatric, are linked to a wide variation in coloration that allows them to be visually recognized and maintain a reproductive isolation. The diversification also affects specializations in diet that may be linked, at least in part, to the phenotypic plasticity of their pharyngeal jaws. This type of dentition, formed by the fusion of the ceratobranchials, has been known since the Eocene-Oligocene boundary in Fayoum, Egypt [MUR 02]. Articulated specimens of cichlids corresponding to five distinct species of the genus *Mahengechromis* in Mahenge, Tanzania, show that the tendency to form species flocks was already present in this family in the Eocene [MUR 01b].

Currently, cyprinodontiforms have a species flock within the genus *Cyprinodon* in Lake Chichancanab in Mexico [HOR 05] and the genus *Orestias* in Lake Titicaca in South America [NOR 00]. A species flock has also been proposed for the extinct aplocheiloid genus *Kenyaichthys* from the Upper Miocene in Kenya [ALT 15].

Mormyrids form another species flock in Africa on the western side of the continent. This is the only species flock in a fluvial environment [SUL 02]. The diversification affects the morphology as well as the nature of the electric signals that these small fishes produce. The scant fossil record of the mormyrids, which are all small fishes with fragile skeletons and minuscule teeth, does not give us any information about the history of such a particular species flock.

Within the salmonids, the the coregonins subfamily has some species flocks in North America and Europe, although they are not as diversified as the tropical species flocks. These radiations followed the development of lakes after the retreat of the glaciers at the end of the Pleistocene [TUR 99]. The diversification affects the trophic adaptations and reproductive isolations. The species complexes present a reticulate evolution with cases of phenotypic convergence, as has been demonstrated for some Ciscoes in North America [TUR 03] and coregones in Europe [ØST 06]. The salmonids began to diversify, and possibly to form small species flocks, before the radiation of coregones in the Holocene, as indicated by the significant diversity of these fishes in deposits dating from the Upper Miocene to the Pleistocene in the northern United States [STE 16]. The general cooling observed in the Neogene was certainly favorable for the diversification of this family.

We are aware of several examples of radiations within specific extinct clades since the Triassic. In some cases, it is difficult to distinguish, based on the fossil record, a species flock type radiation from a radiation that occurred in a large geographical area such as the ones described in the previous section.

The Newark Supergroup in the United States, dating from the Triassic–Jurassic boundary, yielded very rich faunas of ginglymodians, which lived in great lakes reminiscent of the Great Lakes of East Africa today. A good knowledge of the morphology of these species (up to 21 species have been described in the same lake) paired with a good understanding of the paleoecology and temporal framework allowed McCune [MCC 96] to estimate the evolutionary history and the speciation rates in one of these lakes. She demonstrated that certain species were endemic to the lake while others had colonized it from the outside. The formation time for six species is estimated between 5,000 and 8,000 years, which is very close to the values calculated for the current species of freshwater fishes that evolve quickly, notably the cichlids in the African Great Lakes. It is interesting to note that the genera that accompany the species flocks of ginglymodians, which were "modern" neopterygians at that time, are mostly representatives of ancient non-neopterygian groups. Therefore, like various present-day freshwater environments, this environment was both an important site of diversification for some clades and a refuge site for others.

During the Early Cretaceous, the territory of modern China and its neighboring regions (Japan, Korea, Central Asia) was an area of endemism for several clades of freshwater fishes. This concerns in particular the peipiaosteids (acipenseriforms), chuhsiungichtyids (ichthyodectiforms), lycopterids (osteoglossomorphs), sinamiids (halecomorphs) and siyuichthyids (basal teleosteans). These radiations are not exactly synchronous and do not occur in exactly the same place. Given that the fossil record is incomplete, we cannot presume to precisely know the geographical boundaries of the zones of endemism of these Asian families. Therefore, it is possible that in the end, the apparent endemism only corresponds to a lack of data available for a larger geographical area.

The archeomenids and the catervarioliforms may represent some comparable species flocks on Gondwana (Antarctica and Australia for the archeomenids and Africa for the catervarioliforms) in the Jurassic and Lower Cretaceous. However, these groups are still too little known to understand the mechanism and follow the steps of these diversifications.

To my knowledge, the region that corresponds to Europe does not contain any cases of radiations of freshwater species in the Jurassic or Cretaceous. In the Early Cretaceous, when several endemic and diversified primary freshwater clades prospered in Asia, the European deposits like Montsech, Las Hoyas and Bernissart yield species belonging to primarily marine clades such as the macrosemiids, amiids, caturids, ophiopsids, elopids, and chanids, as well as rare species of freshwater pycnodontiforms. This schema corresponds well to faunas composed of vicarious fishes that can be found today in regions of the world without primary freshwater fishes (Australia, New Guinea, large parts of Madagascar and the Philippines). This situation may have persisted in Europe during a part of the Cenozoic, although the fact that several large deposits from this era correspond to environments with variable salinity clouds our interpretation. The presence of ambassids, gonorynchids, mugilids, atheriniforms, moronids, gerreids and gobiids – all families present in marine environments – alongside strictly freshwater taxa, although non-endemic and not very diversified (lepisosteids, amiids), demonstrates the overrepresentation of "vicarious" forms in the Paleogene. In the Neogene, the dominance of the cypriniforms, salmoniforms and esociforms, primary freshwater fishes, lends a strong Asian and northern character to the ichthyofauna. The absence of radiation by freshwater taxa in the Mesozoic and the Paleogene in Europe is probably linked to the fact that this region has a turbulent paleogeographic history that did not permit the establishment of large lakes lasting millions of years that are indispensable to the development of species flocks.

5.3. Lineage depletion, evolutionary stases and refuge zones

5.3.1. The "living fossils"

As Darwin stated in the *Origin of Species*, freshwater environments are favorable for the preservation of taxa that evolve slowly and that correspond, in a certain way, to intermediaries between distinct zoological groups. Darwin gave these organisms the informal name of "living fossils", an expression that is controversial today. The fishes that he included in this category are seven genera of "ganoids" (probably gars, polypterids and the bowfin) and the lepidosiren. Freshwater environments containing "relics" are considered refuge areas. Today, fresh water in North America could be described as such due to the presence of acipenserids, polyodontids, the only species of amiid, episosteids, hiodontids (a basal clade within the osteoglossiforms) and catostomids (a basal clade within the cypriniforms). Freshwater environments already had a refuge role in the Mesozoic as Patterson [PAT 75] noted about the hybodonts (a basal clade of neoselacians), the "paleoniscoids" and, toward the end of the era, the holosteans, aspidorhynchids and "pholidophoroids". Examples of refuges in freshwater can be observed as early as the base of the Triassic, just after the mass extinction at the Permian-Triassic boundary. The continental deposits of these layers contain greater proportions of "paleopterygians" and subholosteans than neopterygians compared to contemporary marine environments [ROM 16]. The genus *Palaeoniscum*, for example, is a genus that survived across the Permian-Triassic boundary. However, all occurrences in the Permian are of marine origin and all occurrences in the Triassic are of continental origin (Middle Triassic in New

South Wales in Australia and Shaanxi in China). The schema is the same for *Acrodus, Challaia, Boreosomus, Hyllingea, Leptogenichthys* and *Myriolepis*. This tendency can also be observed within the Triassic, therefore without a direct link to the Permian-Triassic crisis, for *Gyrolepis, Ptycholepis* and *Dipteronotus*, of which the most ancient representatives are marine and the youngest are freshwater. It should be noted that the "primitive" taxa that survived across the Permian-Triassic boundary in continental environments are not necessarily morphologically stable. They are sometimes locally diversified, as indicated by their high rate of endemism in fresh water in Australia, South Africa, Central Asia and South America during the Triassic. This situation can also be observed in the Middle Jurassic in the Stanleyville Formation assemblage in the Democratic Republic of Congo that contains catervariolids, pleuropholids and other groups of "pholidophoriforms" that belong to rather old lineages for this time, but still diversified. In the Cretaceous, the Jiehol biota is characterized above all by the diversification of new clades for the period, the lycopterids and other basal osteoglossomorphs, as well as by peipiaosteids, which belong to an ancient lineage. Slightly to the west, in central Asia, the last "paleonisciforms", represented by the uighuroniscids and similar forms, have been found in the continental sediments of the Upper Jurassic and the Lower Cretaceous. In the Paleogene, the lakes of the Green River Formation also contain the last representatives of otherwise diversified groups, notably the last ellimmichthyiforms.

5.3.2. Resilience to biological crises

Romano *et al.* [ROM 16] studied the consequences of the extinction event at the end of the Permian for fishes. They observed relatively high generic rates of extinction, about 57–70% for the marine forms and 60–64% for the freshwater forms. Many apex predators were among the marine forms that disappeared. They also observed the transition of some genera from marine to continental environments as the period boundary passed (but this is not a general phenomenon), as well as a decrease in the corporal size of freshwater forms only.

The mass extinction at the Cretaceous–Paleogene boundary is above all known for the disappearance of non-avian dinosaurs, pterosaurs, various groups of marine reptiles and, among invertebrates, ammonoids and rudists. Other groups, including birds, seem to have been greatly affected, although they did not disappear. Bony fishes did not suffer a very substantial extinction but it was selective. Guinot and Cavin [GUI 15b] showed that 14 of the 15 families of actinopterygians that had their last representatives in the Maastrichtian were exclusively marine. These families primarily included piscivorous fishes located at the top of the food chain, as at the Permian-Triassic boundary. None of the families that were exclusively freshwater in the Maastrichtian went extinct at the boundary. This tendency can also be observed at the level of genera, although it is clearer at the level of the families. The only not strictly marine family to disappear is the aspidorhynchids. This family was marine for most of its history, but its last representatives are known in continental environments at the end of the Cretaceous in North America. Mawsoniid coelacanths should be added to the family of freshwater actinopterygians that went extinct. Like for the aspidorhynchids, the last mawsoniids were confined to brackish and freshwater environments in the Cretaceous with the last known representatives in continental environments at the end of the Cretaceous in Europe. The causes proposed to explain the resilience of freshwater fishes compared to marine forms during the mass extinction are the same as those already proposed to explain the selection observed in continental tetrapods. Sheehan and Fastovsky [SHE 92, SHE 93] suggested, based on continental faunas in Montana, that terrestrial vertebrates were more affected than freshwater vertebrates because freshwater invertebrates were often situated in food chains that relied on detritus. This observation naturally applies to freshwater fishes, which explains their ability to make it through the mass extinction event [CAV 02] (Figure 5.2). This hypothesis also applies to marine fishes, whose forms within food chains reliant on plankton disappeared, while the coastal or deep environment species within food chains reliant on detritus from the rivers and the sea surface seem to better weather the catastrophe [CAV 02, CAV 09]. Certain lentic environments such as large lakes also have trophic chains based on the primary production of phytoplankton or zooplankton. It is interesting to note that the two families of not strictly marine bony fishes that disappear at the Cretaceous–Paleogene boundary contain representatives, which have been considered to be potential filter feeding planktivores. The mawsoniids, particularly large forms, may have been filter feeders (cf. 4.1) that would have suffered from the drop in primary production in fresh water (similarly, the giant latimeriid *Megalocoelacanthus*, which may also have been a filter feeder, went extinct in a marine environment). In the aspidorhynchids, a filter feeding diet has been suggested for the Lower Cretaceous genera *Richmondichthys* [BAR 04] and *Vinctifer* [MAI 94] (although this one also fed on larger prey). The last representatives of the aspidorhynchids from the Upper Cretaceous in North America are not well known, but we can hypothesize that they were also filter feeders. All these observations agree with the hypothesis of a temporary break in photosynthetic activity at the origin of the mass extinction at the Cretaceous–Paleogene boundary that provoked a collapse in the food chain based on primary production. The species involved in food chains relying on detritus, on the other hand, overcame with the extinction event with less difficulty. The consequences of significant volcanic activity or the impact of a meteorite, two events that occurred at the end of the Cretaceous, could have released enough dust into the atmosphere to temporarily block photosynthesis. The suddenness of a meteorite impact, however, seems more compatible with the high selectivity of extinctions observed in freshwater fishes.

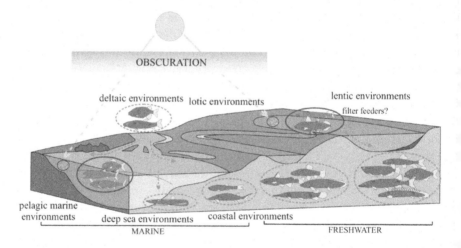

Figure 5.2. *Schema illustrating the consequences of catastrophic events at the Cretaceous–Paleogene boundary on fish faunas. The red silhouettes are the victims and the gray silhouettes are the survivors. Dotted arrows and green lines indicate food chains based on detritus. Red lines indicate food chains based on primary production through photosynthesis (green circles) (according to [CAV 02, CAV 09]). For a color version of the figure, see www.iste.co.uk/cavin/fishes.zip*

Conclusion

The evolutionary history of freshwater fishes in the last 250 million years that is outlined in this book only scratches the surface of what must really have been the true history of this substantial part of continental biodiversity. However, some tendencies can be outlined. A fundamental event in the period under consideration is the profound transformation observed in the composition of fauna between the first part of this period, from the Triassic to the Early Cretaceous, and the second part, which extends from the Late Cretaceous to the present day. For the first part of this history, as of yet we only have vague and poorly structured information. The biogeographical links are difficult to establish because the relations between discovered species are not well known. For the second part of this history, however, we are beginning to understand the broad strokes of the evolution and biogeography of groups of fishes based on fossils. This task has been made easier because the second era corresponds to the appearance of clades of modern freshwater fishes. For these groups, the osteology is much more well known, which makes it possible to better identify the fossil taxa and understand their phylogenetic links. In addition, for a few years now, molecular phylogenetic analyses have provided a set of data that supports, or sometimes challenges, scenarios established based on morphologies. This large quantity of data made it possible to reconstruct the scenarios discussed in this work.

These paleobiogeographical scenarios are rarely simple. They often include both vicariance and dispersion events. Methodological debates about the legitimacy of these two concepts in biogeography have created a very rich scientific literature that is not discussed here. However, it seems that for about two decades now, the goal has no longer been to define the best theoretical concept capable of deciphering a biogeographical history on its own, but rather to find tools that allow us to reconstruct histories that are often complicated and include both vicariance and dispersal events. If the scenarios that we have come across are comparable to stories or narratives, an approach that has been heavily criticized by many biogeographers, it is because the history of life is made up of unique and irreproducible stories. It is up to methodological concepts to adapt to fit the complexity of the living world because the inverse is simply not possible.

The most notable feature in the evolutionary history of freshwater fishes during the last 250 million years of the Earth's history is the extreme diversity of evolutionary dynamics that can be observed. It is in fresh water that we find the fishes, and vertebrates, that have persisted the longest, like the three genera of lungfishes, some acipenseriforms, and even the genus *Esox*. However, it is also in these environments that groups of fishes whose evolutionary dynamic is the most explosive can be found, including, of course, the example of the cichlids, but also the cyprinodontiforms or the atheriniforms. An important observation resulting from this review is that in the Mesozoic, the dual role of freshwater (as a place of both refuge and evolutionary radiation) already existed, but with different actors.

Bibliography

[ADA 10] ADAMSON E.A., HURWOOD D.A., MATHER P.B., "A reappraisal of the evolution of Asian snakehead fishes (Pisces, Channidae) using molecular data from multiple genes and fossil calibration", *Molecular Phylogenetics and Evolution*, vol. 56, pp. 707–717, 2010.

[AGN 10] AGNOLIN F., "A new species of the genus *Atlantoceratodus* (Dipnoiformes: Ceratodontoidei) from the Uppermost Cretaceous of Patagonia and a brief overview of fossil dipnoans from the Cretaceous and Paleogene of South America", *Brazilian Geographical Journal: Geosciences and Humanities Research Medium*, vol. 1, no. 2, pp. 162–210, 2010.

[ALE 13] ALEXANDROU M.A., SWARTZ B.A., MATZKE N.J. *et al.*, "Genome duplication and multiple evolutionary origins of complex migratory behavior in Salmonidae", *Molecular Phylogenetics and Evolution*, vol. 69, pp. 514–523, 2013.

[ALT 15] ALTNER M., REICHENBACHER B., "†*Kenyaichthyidae* fam. nov. and †*Kenyaichthys* gen. nov. First Record of a Fossil Aplocheiloid Killifish (Teleostei, Cyprinodontiformes)", *PLoS One*, vol. 10, no. 4, p. e0123056, 2015.

ALV 13] ALVES Y.M., MACHADO L.P., BERGQVIST L.P. *et al.*, "Redescription of two lungfish (Sarcopterygii: Dipnoi) tooth plates from the late Cretaceous Bauru group, Brazil", *Cretaceous Research*, vol. 40, pp. 243–250, 2013.

AMA 12] AMARAL C.R.L., BRITO P.M., "A new Chanidae (Ostariophysii: Gonorynchiformes) from the Cretaceous of Brazil with affinities to Laurasian Gonorynchiforms from Spain", *PLoS One*, vol. 7, pp. 1–9, 2012.

AMI 10] AMIOT R., WANG X., LÉCUYER C. *et al.*, "Oxygen and carbon isotope compositions of Middle Cretaceous vertebrates from North Africa and Brazil; Ecological and environmental significance", *Palaeogeography, Palaeoclimatology, Palaeoecology*, vol. 297, pp. 439–451, 2010.

[AMI 15] AMIOT R., WANG X., ZHOU Z. *et al.*, "Environment and ecology of East Asian dinosaurs during the Early Cretaceous inferred from stable oxygen and carbon isotopes in apatite", *Journal of Asian Earth Sciences*, vol. 98, pp. 358–370, 2015.

[AND 98] ANDERSON M.E., "A late Cretaceous (Maasteichtian) Galaxiid fish from South Africa", *Special Publication*, vol. 60, pp. 1–8, 1998.

[ANT 90] ANTUNES M.T., MAISEY J.G., MARQUES M.M. *et al.*, *Triassic Fishes from the Cassange Depression (RP de Angola)*, Volumes Especiais, 1990.

[ANT 16] ANTOINE P.-O., ABELLO M.A., ADNET S. *et al.*, "A 60-million-year Cenozoic history of western Amazonian ecosystems in Contamana, eastern Peru", *Gondwana Research*, vol. 31, pp. 30–59, 2016.

[ARA 35] ARAMBOURG C., SCHNEEGANS D., "Poissons fossiles du bassin sédimentaire du Gabon", *Annales de paléontologie*, vol. 24, pp. 138–160, 1935.

[ARA 43] ARAMBOURG C., JOLEAUD L., "Vertébrés fossiles du Bassin du Niger", *Bulletin de la Direction des Mines*, vol. 7, pp. 27–85, 1943.

[ARM 11] ARMBRUSTER J.W., "Global catfish biodiversity", *American Fisheries Society Symposium*, vol. 77, pp. 15–37, 2011.

[ARR 96] ARRATIA G., CIONE A., "The record of Fossil Fishes of Southern South America", *Münchner Geowissenschaftliche Abhandlungen (A)*, vol. 30, pp. 9–72, 1996.

[ARR 97] ARRATIA G., "Basal Teleosts and teleostean phylogeny", *Palaeo Ichthyologica*, vol. 7, pp. 5–168, 1997.

[ARR 99] ARRATIA G., SCHULTZE H.-P., "Semionotiform fish from the Upper Jurassic of Tendaguru (Tanzania). Mitteilungen aus dem Museum für Naturkunde zu Berlin", *Geowissenschaftliche Reihe*, vol. 2, pp. 135–153, 1999.

[ARR 03] ARRATIA G., "Catfish head skeleton–an overview", in ARRATIA G., KAPOOR B.G., CHARDON M. *et al.* (eds), *Catfishes*, vol. 1, Science Publishers, Enfield, 2003.

[ARR 13a] ARRATIA G., "Morphology, taxonomy, and phylogeny of Triassic pholidophorid fishes (Actinopterygii, Teleostei)", *Journal of Vertebrate Paleontology*, vol. 33, pp. 1–138, 2013.

[ARR 13b] ARROYAVE J., DENTON J.S., STIASSNY M.L., "Are characiform fishes Gondwanan in origin? Insights from a time-scaled molecular phylogeny of the Citharinoidei (Ostariophysi: Characiformes)", *PLoS One*, vol. 8, p. e77269, 2013.

[AZP 11] AZPELICUETA M.D.L.M., CIONE A.L., "Redescription of the Eocene catfish *Bachmannia chubutensis* (Teleostei: Bachmanniidae) of southern South America", *Journal of Vertebrate Paleontology*, vol. 31, pp. 258–269, 2011.

[AZP 15] AZPELICUETA M.D.L.M., CIONE A.L., COZZUOL M.A. *et al.*, "*Kooiichthys jono* n. gen. n. sp., a primitive catfish (Teleostei, Siluriformes) from the marine Miocene of southern South America", *Journal of Paleontology*, vol. 89, pp. 791–801, 2015.

[BAR 70] BARDACK, D., "A new teleost from the Oldman Formation (Cretaceous) of Alberta", *National Museums of Canada, Publications in Palaeontology*, vol. 3, pp. 1–8, 1970.

[BAR 73] BARBOUR C.D., "A biogeographical history of Chirostoma (Pisces: Atherinidae): a species flock from the Mexican Plateau", *Copeia*, pp. 533–556, 1973.

[BAR 04] BARTHOLOMAI A., "The large aspidorhynchid fish, *Richmondichthys sweeti* (Etheridge Jnr and Smith Woodward, 1891) from Albian marine deposits of Queensland, Australia", *Memoirs of the Queensland Museum*, vol. 49, no. 2, pp. 521–536, 2004.

[BEA 06] BEAN L., "The leptolepid fish *Cavenderichthys talbragarensis* (Woodward, 1895) from the Talbragar Fish Bed (Late Jurassic) near Gulgong, New South Wales", *Records-Western Australian Museum*, vol. 23, p. 43, 2006.

[BEN 97] BENTON M.J., "Models for the diversification of life", *Trends in Ecology and Evolution*, vol. 12, pp. 490–495, 1997.

[BEN 15] BENYOUCEF M., LÄNG E., CAVIN L. *et al.*, "Overabundance of piscivorous dinosaurs (Theropoda: Spinosauridae) in the mid-Cretaceous of North Africa: the Algerian dilemma", *Cretaceous Research*, vol. 55, pp. 44–55, 2015.

[BET 13] BETANCUR R.R., BROUGHTON R.E., WILEY E.O. *et al.*, "The tree of life and a new classification of bony fishes", *PLoS Currents*, vol. 5, pp. 1–45, 2013.

[BET 14] BETANCUR R.R., WILEY N., BAILLY M. *et al.*, "Phylogenetic classification of bony fishes–version 3", available at: http://www.deepfin.org/Classification_v3.htm, 2014.

[BET 15] BETANCUR R.R., ORTÍ G., PYRON R.A., "Fossil-based comparative analyses reveal ancient marine ancestry erased by extinction in ray-finned fishes", *Ecology Letters*, vol. 18, pp. 441–450, 2015.

[BLA 17] BLANCO A., SZABÓ M., BLANCO-LAPAZ À. et al., "Late Cretaceous (Maastrichtian) Chondrichthyes and Osteichthyes from northeastern Iberia", *Palaeogeography, Palaeoclimatology, Palaeoecology*, vol. 465, pp. 278–294, 2017.

[BOG 13] BOGAN S., TAVERNE L., AGNOLIN F., "First triassic and oldest record of a South American amiiform fish: *Caturus* sp. from the Los Menucos Group (lower Upper Triassic), Rio Negro province, Argentina", *Geologica Belgica*, vol. 16, pp. 191–195, 2013.

[BÖH 03] BÖHME M., REICHENBACHER B., "Teleost fishes from the Karpatian (lower Miocene) of the western paratethys", in BRZOBOHATY R., CICHA I., KOVAC M. et al. (eds), *The Karpatian—An Early Miocene Stage of the Central Paratethys*, Masaryk University, 2003.

[BÖH 04] BÖHME M., "Migration history of air-breathing fishes reveals Neogene atmospheric circulation patterns", *Geology*, vol. 32, pp. 393–396, 2004.

[BON 96] BONDE N., "Osteoglossids (Teleostei: Osteoglossomorpha) of the Mesozoic. Comments on their interrelationships", in ARRATIA G., VIOHLG G. (ed.), *Mesozoic Fishes 1 – Systematics and Paleoecology*, Dr. Friedrich Pfeil, Munich, 1996.

[BON 08] BONDE N., "Osteoglossomorphs of the marine Lower Eocene of Denmark–with remarks on other Eocene taxa and their importance for palaeobiogeography", in CAVIN L., LONGBOTTOM A., RICHTER M. (eds), *Fishes and the Break-up of Pangaea*, Geological Society, London, 2008.

[BOU 02] BOUTON N., WITTE F., VAN ALPHEN J.M.J., "Experimental evidence for adaptive phenotypic plasticity in a rock-dwelling cichlid fish from Lake Victoria", *Biological Journal of the Linnean Society*, vol. 77, pp. 185–192, 2002.

[BRI 03] BRIGGS J.C., "Fishes and birds: Gondwana life rafts reconsidered", *Systematic Biology*, vol. 52, pp. 548–553, 2003.

[BRI 90] BRINKMAN D.B., "Paleooecology of the Judith River Formation (Campanian) of Dinosaur Provincial Park, Alberta, Canada: Evidence from vertebrate microfossil localities", *Palaeogeography, Palaeoclimatology, Palaeoecology*, vol. 78, pp. 37–54, 1990.

[BRI 14a] BRINKMAN D.B., NEWBREY M.G., NEUMAN A.G., "Diversity and paleoecology of actinopterygian fish from vertebrate microfossil localities of the Maastrichtian Hell Creek Formation of Montana", *Geological Society of America Special Papers*, vol. 503, pp. 247–270, 2014.

[BRI 98] BRITO P.M., MARTILL D.M., WENZ S., "A semionotid fish from the Crato Formation (Aptian, Lower Cretaceous) of Brazil: palaeoecological implications", *Oryctos*, vol. 1, pp. 37–42, 1998.

[BRI 02] BRITO P.M., GALLO V., "A new pleuropholid, *Gondwanapleuropholis longimaxillaris* n. g., n. sp. (Actinopterygii: Teleostei) from the Jurassic of north east Brazil", *Comptes Rendus Palévol*, vol. 1, pp. 697–703, 2002.

[BRI 08a] BRITO P.M., ALVARADO-ORTEGA J., "A new species of *Placidichthys* (Halecomorphi: Ionoscopiformes) from the Lower Cretaceous Marizal Formation, northeastern Brazil, with a review of the biogeographical distribution of the Ophiopsidae", in CAVIN L., LONGBOTTOM A., RICHTER M. (eds), *Fishes and the Break-up of Pangaea*, Geological Society, London, 2008.

[BRI 08b] BRITO P.M., YABUMOTO Y., GRANDE L., "New amiid fish (Halecomorphi: Amiiformes) from the Lowesr Cretaceous Crato Formation, Araripe Basin, Northeast Brazil", *Journal of Vertebrate Paleontology*, vol. 28, pp. 1007–1014, 2008.

[BRI 10] BRITO P.M., MEUNIER F., CLÉMENT G. *et al.*, "The histological structure of the calcified lung of the fossil coelacanth *Axelrodichthys araripensis* (Actinistia: Mawsoniidae)", *Palaeontology*, vol. 53, pp. 1281–1290, 2010.

[BRI 14b] BRITZ R., CONWAY K.W., RUEBER L., "Miniatures, morphology and molecules: *Paedocypris* and its phylogenetic position (Teleostei, Cypriniformes)", *Zoological Journal of the Linnean Society*, vol. 172, pp. 556–615, 2014.

[BUR 12] BURRIDGE C. P., MCDOWALL R. M., CRAW D. *et al.*, "Marine dispersal as a pre-requisite for Gondwanan vicariance among elements of the galaxiid fish fauna", *Journal of Biogeography*, vol. 39, no. 2, pp. 306–321, 2012.

[BUS 10] BUSCALIONI A., FREGENAL-MARTÍNEZ M., "A holistic approach to the palaeoecology of Las Hoyas Konservat-Lagerstätte (La Huérguina Formation, Lower Cretaceous, Iberian Ranges, Spain)", *Journal of Iberian Geology*, vol. 36, pp. 297–326, 2010.

[CAM 15] CAMPANELLA D., HUGHES L.C., UNMACK P.J. *et al.*, "Multi-locus fossil-calibrated phylogeny of Atheriniformes (Teleostei, Ovalentaria)", *Molecular Phylogenetics and Evolution*, vol. 86, pp. 8–23, 2015.

[CAM 13] CAMPBELL M.A., LÓPEZ J.A., SADO T. *et al.*, "Pike and salmon as sister taxa: detailed intraclade resolution and divergence time estimation of Esociformes+Salmoniformes based on whole mitochondrial genome sequences", *Gene*, vol. 530, no. 1, pp. 57–65, 2013.

[CAN 11] CANDEIRO C.R.A., FANTI F., THERRIEN F. *et al.*, "Continental fossil vertebrates from the mid-Cretaceous (Albian–Cenomanian) Alcântara Formation, Brazil, and their relationship with contemporaneous faunas from North Africa", *Journal of African Earth Sciences*, vol. 60, pp. 79–92, 2011.

[CAR 03a] CARNEVALE G., SORBINI C., LANDINI W., "*Oreochromis lorenzoi*, a new species of tilapiine cichlid from the late Miocene of central Italy", *Journal of Vertebrate Paleontology*, vol. 23, no. 3, pp. 508–516, 2003.

[CAR 03b] CARPENTER S.J., ERICKSON J.M., HOLLAND F.D., "Migration of a Late Cretaceous fish", *Nature*, vol. 423, no. 6935, pp. 70–74, 2003.

[CAR 08] CARVALHO M.S.S., MAISEY J.G., "New occurrence of *Mawsonia* (Sarcopterygii: Actinistia) from the Early Cretaceous of the Sanfranciscana Basin, Minas Gerais, southeastern Brazil", in CAVIN L., LONGBOTTOM A., RICHTER M. (eds), *Fishes and the Break-up of Pangaea*, Geological Society, London, 2008.

[CAS 61] CASIER E., "Matériaux pour la Faune Ichthyologique Eocrétacique du Congo", *Annales du Musée Royal de l'Afrique Centrale-Tervuren, Belgique, Série 8, Sciences géologiques*, vol. 39, pp. 1–96, 1961.

[CAS 71] CASIER E., TAVERNE L., "Note préliminaire sur le matériel paléoichthyologique éocrétacique récolté par la Spanish Gulf Oil Company en Guinée Equatoriale et au Gabon", *Revue de zoologie et de botanique africaine*, vol. 83, pp. 16–20, 1971.

[CAS 13] CASANE D., LAURENTI P., "Why coelacanths are not 'living fossils'", *Bioessays*, vol. 35, no. 4, pp. 332–338, 2013.

[CAV 91] CAVENDER T. M., "The fossil record of the Cyprinidae", in WINFIELD I.J., NELSON J.S. (eds), *Cyprinid Fishes*, Springer, Dordrecht, 1991.

[CAV 98] CAVENDER T.M., "Development of the North American Tertiary freshwater fish fauna with a look at parallel trends found in the European record", *Italian Journal of Zoology*, vol. 65, pp. 149–161, 1998.

[CAV 96] CAVIN L., MARTIN M., VALENTIN X., "Découverte d'*Atractosteus africanus* (Actinopterygii, Lepisosteidae) dans le Campanien inférieur de Ventabren (Bouches-du-Rhône, France). Implications paléobiogéographiques", *Revue de Paléobiologie*, vol. 15, pp. 1–7, 1996.

[CAV 99a] CAVIN L., "A new Clupavidae (Teleostei, Ostariophysi) from the Cenomanian of Daoura (Morocco)", *Comptes Rendus de l'Académie des Sciences, Sciences de la Terre et des Planètes*, vol. 329, pp. 689–695, 1999.

[CAV 99b] CAVIN L., "Osteichthyes from the Upper Cretaceous of Lano (Iberian Peninsula)", *Estudios del Museo de Ciencias Naturales de Alava*, vol. 14, pp. 105–110, 1999.

[CAV 01a] CAVIN L., BRITO P.M., "A new Lepisosteidae (Actinopterygii: Ginglymodi) from the Cretaceous of the Kem Kem beds, southern Morocco", *Bulletin de la Société géologique de France*, vol. 172, pp. 141–150, 2001.

[CAV 01b] CAVIN L., FOREY P.L., "Osteology and systematic affinities of *Palaeonotopterus greenwoodi* Forey 1997 (Teleostei: Osteoglossomorpha)", *Zoological Journal of the Linnean Society*, vol. 132, pp. 1–28, 2001.

[CAV 03] CAVIN L., SUTEETHORN V., KHANSUBHA S. *et al.*, "A new Semionotid (Actinopterygii, Neopterygii) from the Late Jurassic–Early Cretaceous of Thailand", *Comptes Rendus Palevol*, vol. 2, pp. 291–297, 2003.

[CAV 04] CAVIN L., FOREY P.L., "New mawsoniid coelacanth (Sarcopterygii: Actinistia) remains from the Cretaceous of the Kem Kem beds, SE Morocco", in TINTORI A., ARRATIA G. (eds), *Mesozoic Fishes 3 – Systematics, Plaeoenvironments and Biodiversity*, Dr. Friedrich Pfeil, Munich, 2004.

[CAV 05] CAVIN L., FOREY P.L., BUFFETAUT E. *et al.*, "Latest European coelacanth shows Gondwanan affinities", *Biology Letters*, vol. 1, pp. 176–177, 2005.

[CAV 06] CAVIN L., SUTEETHORN V., "A new Semionotiform (Actinopterygii, Neopterygii) from Upper Jurassic–Lower Cretaceous deposits of North-East Thailand, with comments on the relationships of Semionotiforms", *Palaeontology*, vol. 49, pp. 339–353, 2006.

[CAV 07a] CAVIN L., SUTEETHORN V., BUFFETAUT E. *et al.*, "A new Thai Mesozoic lungfish (Sarcopterygii, Dipnoi) with an insight into post-Palaeozoic dipnoan evolution", *Zoological Journal of the Linnean Society*, vol. 149, pp. 141–177, 2007.

[CAV 07b] CAVIN L., SUTEETHORN V., BUFFETAUT E. *et al.*, "The first sinamiid fish (Holostei, Halecomorpha) from South-east Asia (Early Cretaceous of Thailand)", *Journal of Vertebrate Paleontology*, vol. 27, pp. 827–837, 2007.

[CAV 08a] CAVIN L., "Palaeobiogeography of cretaceous bony fishes (Actinistia, Dipnoi and Actinopterygii)", in CAVIN L., LONGBOTTOM A., RICHTER M. (eds), *Fishes and the Break-up of Pangaea*, Geological Society, London, 2008.

[CAV 08b] CAVIN L., FOREY P.L., "A new tselfatiiform teleost from the mid-Cretaceous (Cenomanian) of the Kem Kem beds, Southern Morocco", in ARRATIA G., SCHULTZE H.-P., WILSON M.V.H. (eds), *Mesozoic Fishes 4 – Homology and Phylogeny*, Dr. Friedrich Pfeil, Munich, 2008.

[CAV 09] CAVIN L., DEESRI U., SUTEETHORN V., "The Jurassic and Cretaceous bony fish record (Actinopterygii, Dipnoi) from Thailand)", in BUFFETAUT E., CUNY G., LE LOEUFF J. *et al.* (eds), *Late Palaeozoic and Mesozoic Continental Ecosystems in SE Asia*, Geological Society, London, pp. 125–139, 2009.

[CAV 10a] CAVIN L., "On giant filter feeders", *Science*, vol. 327, no. 5968, pp. 968–969, 2010.

[CAV 10b] CAVIN L., "Diversity of Mesozoic semionotiform fishes and the origin of gars (Lepisosteidae)", *Naturwissenschaften*, vol. 97, pp. 1035–1040, 2010.

[CAV 10c] CAVIN L., TONG H., BOUDAD L. *et al.*, "Vertebrate assemblages from the early Late Cretaceous of southeastern Morocco: an overview", *Journal of African Earth Sciences*, vol. 57, pp. 391–412, 2010.

[CAV 11] CAVIN L., KEMP A., "The impact of fossils on the evolutionary distinctiveness and conservation status of the Australian lungfish", *Biological Conservation*, vol. 144, pp. 3140–3142, 2011.

[CAV 12] CAVIN L., GINER S., "A large halecomorph fish (Actinopterygii: Holostei) from the Valanginian (Early Cretaceous) of southeast France", *Cretaceous Research*, vol. 37, pp. 201–208, 2012.

[CAV 13a] CAVIN L., DEESRI U., SUTEETHORN V., "Osteology and relationships of *Thaiichthys* nov. gen.: a Ginglymodi from the Late Jurassic–Early Cretaceous of Thailand", *Palaeontology*, vol. 56, pp. 183–208, 2013.

[CAV 13b] CAVIN L., FURRER H., OBRIST C., "New coelacanth material from the Middle Triassic of eastern Switzerland, and comments on the taxic diversity of actinistans", *Swiss Journal of Geosciences*, vol. 106, pp. 161–177, 2013.

[CAV 13c] CAVIN L., FOREY P.L., GIERSH S., "Osteology of *Eubiodectes libanicus* (Pictet & Humbert, 1866) and some other ichthyodectiformes (Teleostei): phylogenetic implications", *Journal of Systematic Palaeontology*, vol. 11, pp. 113–175, 2013.

[CAV 14a] CAVIN L., DEESRI U., SUTEETHORN V., "Ginglymodian fishes (Actinopterygii, Holostei) from Thailand: an overview", *Journal of Science and Technology*, vol. 33, pp. 348–356, 2014.

[CAV 14b] CAVIN L., GUINOT G., "Coelacanths as 'almost living fossils'", *Frontiers in Ecology and Evolution*, vol. 2, pp. 1–5, 2014.

[CAV 15] CAVIN L., BOUDAD L., TONG H. *et al.*, "Taxonomic composition and trophic structure of the continental bony fish assemblage from the early Late Cretaceous of southeastern Morocco", *Plos One*, vol. 10, p. e0125786, 2015.

[CAV 16] CAVIN L., VALENTIN X., GARCIA G., "A new mawsoniid coelacanth (Actinistia) from the Upper Cretaceous of Southern France", *Cretaceous Research*, vol. 62, pp. 65–73, 2016.

[CHA 96] CHANG M.-M., JIN F., "Mesozoic fish faunas of China", in ARRATIA G., VIOHL G. (eds), *Mesozoic Fishes 1 – Systematics and Paleoecology*, Dr. Friedrich Pfeil, Munich, 1996.

[CHA 97] CHANG M.-M., GRANDE L., Redesription of *Paraclupea chetungensis*, an Early Clupeomorph from the Lower Cretaceous of Southeastern China", *Fieldiana, Geology, New Series*, vol. 7, pp. 1–19, 1997.

[CHA 01] CHANG M.-M., MIAO D., CHEN Y. *et al.*, "Suckers (Fish, Catosomidae) from the Eocene of China account for the family's current disjunct distribution", *Science in China (Series D)*, vol. 44, pp. 577–586, 2001.

[CHA 03] CHARDON M., PARMENTIER E., VANDEWALLE P., "Morphology, development and evolution of the Weberian apparatus in catfish", *Catfishes*, vol. 1, pp. 71–120, 2003.

[CHA 04] CHANG M.M., MIAO D., "An overview of Mesozoic fishes in Asia", in ARRATIA G., TINTORI A. (eds), *Systematics, Paleoenvironments and Biodiversity*, Dr. Friedrich Pfeil, Serpiano, 2004.

[CHA 08] CHANG M.-M., CHEN G., "Fossil Cypriniformes from China and its adjacent areas and their palaeobiogeographical implications", in CAVIN L., LONG BOTTOM A., RICHTER H. (eds), *Fishes and the Break-up of Pangaea*, Geological Society, London, 2008.

[CHA 10] CHANG M.-M., WANG N., WU F.-X., "Discovery of †*Cyclurus* (Amiinae, Amiidae, Amiiformes, Pisces) from China", *Vertebrata PalAsiatica*, vol. 48, pp. 85–100, 2010.

[CHE 88] CHEN P., "Distribution and migration of the Jehol Fauna with reference to non-marine Jurassic-Cretaceous boundary in China", *Acta Palaeontologica Sinica*, vol. 27, pp. 659–683, 1988.

[CHE 13] CHEN W.-J., LAVOUÉ S., MAYDEN R.L., "Evolutionary origin and early biogeography of otophysan fishes (Ostariophysi: Teleostei)", *Evolution*, vol. 67, pp. 2218–2239, 2013.

[CHE 14] CHEN W.J., LAVOUÉ S., BEHEREGARAY L.B. *et al.*, "Historical biogeography of a new antitropical clade of temperate freshwater fishes", *Journal of Biogeography*, vol. 41, pp. 1806–1818, 2014.

[CHI 98] CHIAPPE L., RIVARIOLA D., CIONE A. *et al.*, "Biotic association and paleoenvironmental reconstruction of the 'Loma del Pterodaustro' fossil site (Early Cretaceous, Argentina)", *Geobios*, vol. 31, pp. 349–369, 1998.

[CHO 52] CHOUBERT G., CLARIOND L., HINDERMEYER J., "Livret-guide de l'excursion C 36. Anti-Atlas central et oriental., no. 11. Rabat", *Congrès géologique international, XIXe session (Alger 1952), série: Maroc, Morocco, Africa, Cretaceous, geology, sedimentology*, 1952.

[CHU 06] CHURCHER C.S., DE IULIIS G., KLEINDIENST M.R., "A new genus for the Dipnoan species *Ceratodus tuberculatus* Tabaste 1963", *Geodiversitas*, vol. 28, no. 4, pp. 635–647, 2006.

[CIO 80] CIONE A.L., LAFITTE G., "El primer siluriforme (Pisces, Ostariophysi) del Cretácico de Patagonia. Consideraciones sobre el área de la diferenciación de los Siluriformes. Aspectos biogeográficos", *Segundo Congreso Argentino de Paleontología y Bioestratigrafía y Primer Congreso Latinoamericano de Paleontología*, Actas, Buenos Aires, pp. 35–48, 1980.

[CIO 85] CIONE A.L., PEREIRA S.M., ALONSO R. *et al.*, "Los bagres (Osteichthyes, Siluriformes) de la Formacion Yacoraite (Cretacico tardio) del Noroeste Argentino: Consideraciones biogeographificas y bioestratigraficas", *Ameghiniana*, vol. 21, no. 2, pp. 294–304, 1985.

[CIO 87] CIONE A.L., "The Late Cretaceous fauna of Los Alamitos, Patagonia, Argentina. Part II–The Fishes" *Revista del Museo argentino de Ciencias Naturales "Bernardino Rivadavia" e Instituto nacional de Investigacion de Las Ciencias Naturales*, vol. 3, pp. 111–120, 1987.

[CIO 02] CIONE A.L., PRASAD G.V.R., "The oldest known catfish (Teleostei: Siluriformes) from Asia (India, Late Cretaceous)", *Journal of Paleontology*, vol. 76, pp. 190–193, 2002.

[CIO 11] CIONE A.L., GOUIRIC-CAVALLI S., GELFO J.N. *et al.*, "The youngest non-lepidosirenid lungfish of South America (Dipnoi, latest Paleocene–earliest Eocene, Argentina)", *Alcheringa*, vol. 35, no. 2, pp. 193–198, 2011.

[CIO 12] CIONE A.L., GOUIRIC-CAVALLI S., "*Metaceratodus kaopen* comb. nov. and *M. wichmanni* comb. nov., two Late Cretaceous South American species of an austral lungfish genus (Dipnoi)", *Alcheringa*, vol. 36, no. 2, pp. 203–216, 2012.

[CLO 91] CLOUTIER R., "Patterns, trends, and rates of evolution within the Actinistia", *Environmental Biology of Fishes*, vol. 32, pp. 23–58, 1991.

[COS 11] COSTA W.J.E.M., "Redescription and phylogenetic position of the fossil killifish †*Carrionellus diumortuus* White from the Lower Miocene of Ecuador (Teleostei: Cyprinodontiformes)", *Cybium*, vol. 35, pp. 181–187, 2011.

[COS 12] COSTA W.J.E.M., "Oligocene killifishes (Teleostei: Cyprinodontiformes) from southern France: relationships, taxonomic position, and evidence of internal fertilization", *Vertebrate Zoology*, vol. 62, pp. 371–386, 2012.

[CRO 74] CROIZAT L., NELSON G., ROSEN D.E., "Centers of origin and related concepts", *Systematic Biology*, vol. 23, no. 2, pp. 265–287, 1974.

[CUP 15] CUPELLO C., BRITO P.M., HERBIN M. et al., "Allometric growth in the extant coelacanth lung during ontogenetic development", Nature Communications, vol. 6, pp. 1–5, 2015.

[DAV 13] DAVIS M.P., ARRATIA G., KAISER T.M., "The first fossil shellear and its implications for the evolution and divergence of the Kneriidae (Teleostei: Gonorynchiformes)", in ARRATIA G., SCHULTZE H.-P., WILSON M.V.H. (eds), Mesozoic Fishes 5 – Global Diversity and Evolution, Dr. Friedrich Pfeil, Munich, 2013.

[DAV 99] DAVIS S.P., MARTILL D.M., "The gonorynchiform fish Dastilbe from the Lower Cretaceous of Brazil", Palaeontology, vol. 42, pp. 715–740, 1999.

[DEE 09] DEESRI U., CAVIN L., CLAUDE J. et al., "Morphometric and taphonomic study of a ray-finned fish assemblage (Lepidotes buddhabutrensis, Semionotidae) from tha Late Jurassic-Early Cretaceous of NE Thailand", in BUFFETAUT E., CUNY G., LE LOEUFF J. et al. (eds), Late Palaeozoic and Mesozoic Continental Ecosystems in SE Asia, Geological Society, London, 2009.

[DEE 14] DEESRI U., LAUPRASERT K., SUTEETHORN V. et al., "A New Ginglymodian fish (Actinopterygii, Holostei) from the Late-Jurassic Phu Kradung Formation, northeastern Thailand", Acta Palaeontologia Polonica, vol. 59 pp. 313–331, 2014.

[DEE 16] DEESRI U., JINTASAKUL P., CAVIN L., "A new Ginglymodi (Actinopterygii, Holostei) from the Late Jurassic–Early Cretaceous of Thailand, with comments on the early diversification of Lepisosteiformes in Southeast Asia", Journal of Vertebrate Paleontology, vol. 36, no. 6, p. e1225747, 2016.

[DEN 41] DENISON R.H., "The soft anatomy of Bothriolepis", Journal of Paleontology, pp. vol. 15, 553–561, 1941.

[DEQ 05] DE QUEIROZ A., "The resurrection of oceanic dispersal in historical biogeography", Trends in Ecology and Evolution, vol. 20, pp. 68–73, 2005.

[DIL 11] DILLMAN C.B., BERGSTROM D.E., NOLTIE D.B. et al., "Regressive progression, progressive regression or neither? Phylogeny and evolution of the Percopsiformes (Teleostei, Paracanthopterygii)", Zoologica Scripta, vol. 40, pp. 45–60, 2011.

[DIO 04] DIOGO R., "Phylogeny, origin and biogeography of catfishes: support for a Pangean origin of 'modern teleosts' and reexamination of some Mesozoic Pangean connections between Gondwanan and Laurasian supercontinents", Animal Biology, vol. 54, pp. 331–351, 2004.

[DIV 15] DIVAY J.D., MURRAY A.M., "The late Eocene–early Oligocene ichthyofauna from the Eastend area of the Cypress Hills Formation, Saskatchewan, Canada", *Journal of Vertebrate Paleontology*, vol. 35, p. e956877, 2015.

[DIV 16] DIVAY J.D., MURRAY A.M., "An early Eocene fish fauna from the Bitter Creek area of the Wasatch Formation of southwestern Wyoming, USA", *Journal of Vertebrate Paleontology*, vol. 36, no. 6, p. e1196211, 2016.

[DUT 14] DUTEL H., PENNETIER E., PENNETIER G., "A giant marine coelacanth from the Jurassic of Normandy, France", *Journal of Vertebrate Paleontology*, vol. 4, pp. 1239–1242, 2014.

[DUT 15] DUTEL H., HERBIN M., CLÉMENT G., "First occurrence of a mawsoniid coelacanth in the Early Jurassic of Europe", *Journal of Vertebrate Paleontology*, vol. 35, no. 3, p. e929581, 2015.

[DUT 99a] DUTHEIL D.B., "The first articulated fossil cladistian: *Serenoichthys kemkemensis*, gen. et sp. nov., from the Cretaceous of Morocco", *Journal of Vertebrate Paleontology*, vol. 19, pp. 243–246, 1999.

[DUT 99b] DUTHEIL D.B., "An overview of the freshwater fish fauna from the Kem Kem beds (Late Cretaceous: Cenomanian) of southeastern Morocco", in ARRATIA G., SCHULTZE H.-P. (eds), *Mesozoic Fishes 2 – Systematics and Fossil Record*, Dr. Friedrich Pfeil, Munich, 1999.

[DZI 80] DZIEWA T.J., "Early Triassic osteichthyans from the Knocklofty Formation of Tasmania", *Papers and Proceedings of the Royal Society of Tasmania*, pp. 145–160, 1980.

[EBE 16] EBERT M., "The Pycnodontidae (Actinopterygii) in the late Jurassic: 2) *Turboscinetes* gen. nov. in the Solnhofen Archipelago (Germany) and Cerin (France)", *Archaeopteryx* vol. 33, pp. 12–53, 2016.

[EST 69] ESTES R., "Studies on fossil Phyllodont fishes: interrelationships and evolution in the Phyllodontidae (Albuloidei)", *Copeia*, vol. 2, pp. 317–331, 1969.

[EST 78] ESTES R., HIATT R., "Studies on fossil phyllodont fishes: a new species of *Phyllodus* (Elopiformes, Albuloidea) from the Late Cretaceous of Montana", *Paleobios*, vol. 28, pp. 1–10, 1978.

[ETT 04] ETTACHFINI E.M., ANDREU B., "Le Cénomanien et le Turonien de la Plate-forme Préafricaine du Maroc", *Cretaceous Research*, vol. 25, pp. 277–302, 2004.

[FAR 07] FARA E., GAYET M., TAVERNE L., "Les Gonorynchiformes fossiles: distribution et diversité", *Cybium*, vol. 31, pp. 115–122, 2007.

[FAR 10] FARA E., GAYET M., TAVERNE L., "The fossil record of Gonorynchiformes", in GRANDE T., POYATO-ARIZA F.R.D. (eds), *Gonorynchiformes and Ostariophysan Relationships: A Comprehensive Review*, Science Publishers, Enfield, 2010.

[FER 07] FERRARIS C.J., *Checklist of catfishes, recent and fossil (Osteichthyes: Siluriformes), and catalogue of siluriform primary types*, Magnolia Press, Auckland, 2007.

[FIG 05] FIGUEIREDO F.J., "Reassessment of the morphology of *Scombroclupeoides scutata* Woodward, 1908, a Teleostean fish from the early Cretaceous of Bahia, with comments on its relationships", *Arquivos du Museu Nacional Rio de Janeiro*, vol. 63, pp. 507–522, 2005.

[FIL 01] FILLEUL A., DUTHEIL D.B., "*Spinocaudichthys oumtkoutensis*, a freshwater acanthomorph from the Cenomanian of Morocco", *Journal of Vertebrate Paleontology*, vol. 21, pp. 774–780, 2001.

[FIL 04a] FILLEUL A., DUTHEIL D.B., "A peculiar diplospondylous actinopterygian fish from the Cretaceous of Morocco", *Journal of Vertebrate Paleontology*, vol. 24, pp. 290–298, 2004.

[FIL 04b] FILLEUL A., MAISEY J.G., "Redescription of *Santanichthys diasii* (Otophysi, Characiformes) from the Albian of the Santana Formation and comments on its implications for Otophysan relationships", *American Museum Novitates*, no. 3455, pp. 1–21, 2004.

[FIN 81] FINK S.V., FINK W.L., "Interrlationships of the ostariopysan fishes (Teleostei)", *Zoological Journal of the Linnean Society*, vol. 72, pp. 297–353, 1981.

[FIN 96] FINK S.V., FINK W.L., "Interrelationships of Ostariophysan fishes (Teleostei)", in STIASSNY M.L.J., PARENTI L.R., JOHNSON G.D. (eds), *Interrelationships of Fishes*, Academic Press, San Diego, 1996.

[FOR 73] FOREY P., GARDINER B., "A new dictyopygid from the Cave Sandstone of Lesotho, southern Africa", *Palaeontologia Africana*, vol. 15, pp. 29–31, 1973.

[FOR 97] FOREY P.L., "A Cretaceous notopterid (Pisces: Osteoglossomorpha) from Morocco", *South African Journal of Science*, vol. 93, pp. 564–569, 1997.

[FOR 98] FOREY P.L., *History of the Coelacanth Fishes*, Chapman and Hall, London, 1998.

[FOR 06] FOREY P.L., PATTERSON C., "Description and systematic relationships of *Tomognathus*, an enigmatic fish from the English Chalk", *Journal of Systematic Palaeontology*, vol. 4, pp. 157–184, 2006.

[FOR 07] FOREY P.L., CAVIN L., "A new species of *Cladocyclus* (Teleostei: Ichthyodectiformes) from the Cenomanian of Morocco", *Palaeontologia Electronica*, vol. 10, no. 3, pp. 10, 2007.

[FOR 10] FOREY P.L., HILTON E.J., "Two new Tertiary osteoglossid fishes (Teleostei: Osteoglossomorpha) with notes on the history of the family", in ELLIOTT D.K., MAISEY J.G., YU X. *et al.* (eds), *Morphology, Phylogeny and Paleobiogeography of Fossil Fishes*, Dr. Friedrich Pfeil, Munich, 2010.

[FOR 16] FOREY P.L., "Smith Woodward's ideas on fish classification", *Geological Society, London, Special Publications*, vol. 430, pp. 115–127, 2016.

[FRI 03] FRIEDMAN M., TARDUNO J.A., BRINKMAN D.B., "Fossil fishes from the high Canadian Arctic: further palaeobiological evidence for extreme climatic warmth during the Late Cretaceous (Turonian–Coniacian)", *Cretaceous Research*, vol. 24, pp. 615–632, 2003.

[FRI 06] FRIEDMAN M., COATES M.I., "A new recognized fossil coelacanth highlights the early morphological diversification of the clade", *Proceedings of the Royal Society B*, vol. 273, pp. 245–250, 2006.

[FRI 13] FRIEDMAN M., KECK B.P., DORNBURG A. *et al.*, "Molecular and fossil evidence place the origin of cichlid fishes long after Gondwanan rifting", *Proceedings of the Royal Society B*, vol. 280, pp. 1–8, 2013.

[GAU 75] GAUDANT J., "Intérêt paléoécologique de la découverte de *Gobiu aries* (AG.) (Poisson téléostéen, Gobioidei) dans l'Oligocène des bassins de Marseille et de Saint-Pierre.lès-Martiques (Bouches-du-Rhône)", *Géologie méditerranéenne*, vol. 2, pp. 111–114, 1975.

[GAU 77] GAUDANT J., "Nouvelles observations sur l'ichthyofaune stampienne d'Oberdorf (Canton de Soleure)", *Eclogae Geologicae Helvetiae*, vol. 70, pp. 789–809, 1977.

[GAU 78a] GAUDANT J., "Découverte du plus ancien représentant connu du genre *Esox* L. (Poisson téléostéen, Esocoidei) dans le Stampien moyen du bassin d'Apt (Vaucluse)", *Géologie Méditerranéenne V*, vol. 5, pp. 257–268, 1978.

[GAU 78b] GAUDANT J., "Sur les conditions de gisement de l'ichthyofaune oligocène d'Aix-en-Provence (Bouches-du-Rhône): essai de définition d'un modèle paléoécologique et paléogéographique", *Géobios*, vol. 11, pp. 393–397, 1978.

[GAU 79a] GAUDANT J., "Mise au point sur l'ichthyofaune paléocène de Menat (Puy-de-Dôme)", *Comptes Rendus de l'Académie des Sciences*, vol. 288, pp. 1461–1464, 1979.

[GAU 79b] GAUDANT J., "'*Pachylebias*' *crassicaudus* (Agassiz) (poissons téléostéen, Cyprinodontiforme), un constituant majeur de l'ichthyofaune du Messinien continental du bassin méditerranéen", *Geobios*, vol. 12, pp. 47–73, 1979.

[GAU 79c] GAUDANT J., ROUSSET C., "Découverte de restes de Cyprinidae (Poissons téléostéens) dans l'Oligocène moyen de Marseille (Bouches-du-Rhône)", *Géobios*, vol. 12, pp. 331–337, 1979.

[GAU 81a] GAUDANT J., "Mise au point sur l'ichthyofaune oligocène des anciennes plâtrières d'Aix-en-Provence (Bouches-du-Rhône)", *Comptes Rendus de l'Académie des Sciences, Paris, série III*, vol. 292, pp. 1109–1112, 1981.

[GAU 81b] GAUDANT J., "Nouvelles recherches sur l'ichthyofaune des gypses et des marnes supragypseuses (Eocène supérieur) des envisons de Paris", *Bulletin du B.R.G.M. section IV*, vol. 2, no. 1, pp. 57–75, 1981.

[GAU 82] GAUDANT J., "Apport de l'ichthyofaune à la caractérisation des milieux saumâtres cénozoïques", *Mémoire de la Société Géologique de France (N.S.)*, vol. 144, pp. 139–146, 1982.

[GAU 84a] GAUDANT J., "Nouvelles recherches sur les Cyprinidae (Poissons téléostéens) oligocènes des Limagnes", *Géobios*, vol. 17, pp. 659–666, 1984.

[GAU 84b] GAUDANT J., "Sur la présence de "Percichthyidae" (Poissons téléostéens) dans l'Eocène moyen du Bassin du Duero (Province de Zamora, Espagne)", *Acta Geologica Hispanica*, vol. 19, pp. 139–142, 1984.

[GAU 84c] GAUDANT J., BURKHARDT T., "Sur la découverte de poissons fossiles dans les marnes grises rayées de la zone fossilifère (Oligocène basal) d'Altkirch (Haut-Rhin)", *Sciences géologiques, Bulletin*, vol. 37, pp. 153–171, 1984.

[GAU 85] GAUDANT J., "Mise en évidence d'Osmeridae (Poissons téléostéens, Salmoniformes) dans l'Oligocène lacustre d'Europe occidentale", *Comptes Rendus de l'Académie des Sciences, Paris, série II*, vol. 300, pp. 79–82, 1985.

[GAU 87] GAUDANT J., "Sur la présence de Chandidae (Poissons téléostéens, Percoidei) dans le Cénozoïque européen", *Comptes Rendus de l'Académie des Sciences, Paris, série II*, vol. 304, pp. 1249–1252, 1987.

[GAU 88] GAUDANT J., "Mise au point sur l'ichthyofaune oligocène de Rott, Stösschen et Orsberg (Allemagne)", *Comptes Rendus de l'Académie des Sciences, Paris, série II*, vol. 306, pp. 831–834, 1988.

[GAU 89] GAUDANT J., "L'ichthyofaune stampienne des environs de Chartres-de-Bretagne, près de Rennes (Ille-et-Vilaine): un réexamen", *Géologie de la France*, vol. 1–2, pp. 41–54, 1989.

[GAU 02] GAUDANT J., WEIDMANN M., BERGER J.P. *et al.*, "Recherches sur les dents pharyngiennes de Poissons Cyprinidae de la Molasse d'eau douce oligo-miocène de Suisse (USM, OSM) et de Haute-Savoie (France)", *Revue de Paléobiologie*, vol. 21, pp. 371–389, 2002.

[GAU 05] GAUDANT J., REICHENBACHER B., *"Hemitrichas stapfififi* n. sp. (Teleostei, Atherinidae) with otoliths in situ from the late Oligocene of the Mainz Basin", *Zitteliana* , pp. 189–198, 2005.

[GAU 07] GAUDANT J., "Occurrence of the genus *Tarsichthys* Troschel (Teleostean fishes, Cyprinidae) in the Upper Oligocene of Lake Kunkskopf, near Burgbrohl (E-Eifel-Mountains, Germany)", *Zitteliana*, pp. 127–132, 2007.

[GAU 12a] GAUDANT J., "An attempt at the palaeontological history of the European mudminnows (Pisces, Teleostei, Umbridae)", *Neues Jahrbuch für Geologie und Paläontologie-Abhandlungen*, vol. 263, no. 2, pp. 93–109, 2012.

[GAU 12b] GAUDANT J., SCHAAL S.F., WEI S., "A short account on the Eocene fish fauna from Huadian (Jilin Province, China)", *Palaeobiodiversity and Palaeoenvironments*, vol. 92, pp. 417–423, 2012.

[GAU 13] GAUDANT J., "Présence d'un Osmeridae: *Enoplophthalmus schlumbergeri* Sauvage, 1880 dans l'Oligocène inférieur des environs de Céreste (Alpes-de- Haute-Provence, France)", *Geodiversitas*, vol. 35, pp. 345–357, 2013.

[GAU 15] GAUDANT J., CARNEVALE G., *"Pharisatichthys aquensis* n. gen., n. sp.: un nouveau poisson fossile (Teleostei, Gerreidae) de l'Oligocène supérieur d'Aix-en-Provence (Bouches-du-Rhône, France)", *Geodiversitas*, vol. 37, pp. 109–118, 2015.

[GAU 16] GAUDANT J., *"Francolebias arvernensis* n. sp., une nouvelle espèce de poissons cyprinodontiformes oligocènes de Chadrat (Saint-Saturnin, Puy-de-Dôme, France), avec une brève notice sur un Umbridae fossile du même gisement", *Geodiversitas*, vol. 38, no. 3, pp. 435–449, 2016.

[GAY 81] GAYET M., "Contribution à l'étude anatomique et systématique de l'ichthyofaune cénomanienne du Portugal. Deuxième partie: Les Ostariophysaires", *Communicações dos Serviços Geológicos de Portugal*, vol. 67, pp. 173–190, 1981.

[GAY 85] GAYET M., "Rôle de l'évolution de l'appareil de Weber dans la phylogénie des Ostariophysi suggéré par un nouveau Characiforme du Cénomanien supérieur marin du Portugal", *Comptes Rendus de l'Académie des Sciences, Paris, série II*, vol. 300, no. 17, pp. 895–898, 1985.

[GAY 86] GAYET M., MEUNIER F., "Apport de l'étude de l'ornementation microscopique de la ganoïne dans la détermination de l'appartenance générique et/ou spécifique des écailles isolées", *Comptes Rendus de l'Académie des Sciences, Paris, Série II*, vol. 303, no. 13, pp. 1259–1262, 1986.

[GAY 87] GAYET M., "Lower vertebrates from the early-middle Eocene Kuldana Formation of Kohat (Pakistan): Holostei and Teleostei", *Contributions from the Museum of Paleontology, University of Michigan*, vol. 27, no. 7, pp. 151–168, 1987.

[GAY 88a] GAYET M., "*Gharbouria libanica* nov. gen., nov. sp., "Salmoniforme" nouveau en provenance d'Aïn-el-Ghârboûr, nouveau gisement cénomanien du Liban", *Bulletin du Muséum national d'histoire naturelle. Section C, Sciences de la terre, paléontologie, géologie, minéralogie*, sec. 3, vol. 10, pp. 199–225, 1988.

[GAY 88b] GAYET M., "Le plus ancien crâne de Siluriorme: *Andinichthys bolivianensis* nov. gen., nov. sp. (Andinicthyidae nov. fam.) du Maastrichtien de Tiupampa (Bolivia)", *Comptes Rendus de l'Académie des Sciences, Série II*, vol. 307, no. 7, pp. 833–836, 1988.

[GAY 88c] GAYET M., "Découverte du plus ancien Channiforme (Pisces, Teleostei): *Parachannichthys ramnagarensis* n. g., n. sp., dans le Miocène moyen des Siwaliks (Ramnagar, Jammu et Cachemire, Inde): implications paléobiogéographiques", *Comptes rendus de l'Académie des Sciences, Série II, Mécanique, Physique, Chimie, Sciences de l'univers, Sciences de la Terre*, vol. 307, no. 8, pp. 1033–1036, 1988.

[GAY 88d] GAYET M., MEUNIER F.J., Levrat-CLAVIAC V., "Mise en évidence du plus ancien Polypteridae dans le gisement sénonien d'In Becetem (Niger)", *Comptes Rendus de l'Académie des Sciences, Paris, Série II*, vol. 307, pp. 205–210, 1988.

[GAY 89a] GAYET M., BRITO P.M., "Ichtyofaune nouvelle du Crétacé supérieur du groupe Bauru (états de Sao Paulo et Minas Gerais, Brésil)", *Géobios*, vol. 22, pp. 841–847, 1989.

[GAY 89b] GAYET M., "Note préliminaire sur le matériel paléoichthyologique éocrétacique du Rio Benito (sud de Bata, Guinée Equatoriale)", *Bulletin du Muséum National d'histoire Naturelle. Section C, Sciences de la terre, paléontologie, géologie, minéralogie*, vol. 11, pp. 21–31, 1989.

[GAY 90] GAYET M., "Nouveaux Siluriformes du Maastrichtien de Tiupampa (Bolivia)", *Comptes Rendus de l'Académie des Sciences, Série II*, vol. 310, pp. 867–872, 1990.

[GAY 91a] GAYET M., "'Holostean' and Teleostean fishes of Bolivia", in SUAREZ-SORUCO R. (ed.), *Fosiles y facies de Bolivia 1 – Vertebrados*, Revista Técnica YPFB, Santa Cruz, 1991.

[GAY 91b] GAYET M., MEUNIER F., "Polypteridae (Pisces, Cladistia, Polypteriformes) from the Late Cretaceous and Early Paleocene of Bolivia", in SUAREZ-SORUCO R. (ed.), *Fosiles y facies de Bolivia 1 – Vertebrados*, Revista Técnica YPFB, Santa Cruz, 1991.

[GAY 91c] GAYET M., MEUNIER F., "Première découverte de Gymnotiformes fossiles (Pisces, Ostariophysi) dans le Miocène supérieur de Bolivie", *Comptes Rendus de l'Académie des Sciences, Paris, série II*, vol. 313, pp. 471–476, 1991.

[GAY 92a] GAYET M., MEUNIER F., "Polyptériformes (Pisces, Cladistia) du Maastrichtien et du Paléocène de Bolivia", *Geobios M. S.*, vol. 14, pp. 159–168, 1992.

[GAY 92b] GAYET M., RAGE J.-C., SEMPERE T. *et al.*, "Modalités des échanges de vertébrés continentaux entre l'Amérique du Nord et l'Amérique du Sud au Crétacé supérieur et au Paléocène", *Bulletin de la Société géologique de France*, vol. 163, pp. 781–791, 1992.

[GAY 93] GAYET M., SEMPERE T., CAPPETTA H. *et al.*, "La présence de fossiles marins dans le Crétacé terminal des Andes centrales et ses conséquences paléogéographiques", *Palaeogeography, Palaeoclimatology, Palaeoecology*, vol. 102, pp. 283–319, 1993.

[GAY 97] GAYET M., MEUNIER F., WERNER C., "Strange Polypteriformes from the Upper Cretaceous of in Becetem (Niger) and Wadi Milk Formation (Sudan)", *Geobios M.S.*, vol. 20, pp. 249–255, 1997.

[GAY 98] GAYET M., MEUNIER F., "Maastrichtian to Early Paleocene Freshwater Osteichthyes of Bolivia: additions and comments", in MALBARA L.R., REIS R.E., VARI R.P. *et al.* (eds), *Phylogeny and Classification of Neotropical Fishes*, EDIPUCRS, Porto Alegre, 1998.

[GAY 99] GAYET M., OTERO O., "Analyses de la paléodiversification des Siluriformes (Osteichthyes, Teleostei, Ostariophysi)", *Géobios*, vol. 32, pp. 235–246, 1999.

[GAY 01] GAYET M., MEUNIER F., "A propos du genre *Paralepidosteus* (Ginglymodi, Lepisosteidae) du Crétacé gondwanien", *Cybium*, vol. 25, pp. 153–159, 2001.

[GAY 02] GAYET, M., MEUNIER, F., WERNER, C., "Diversification in Polypteriformes and special comparison with the Lepisosteiformes", *Palaeontology*, vol. 45, pp. 361–376, 2002.

[GAY 03] GAYET M., MEUNIER F.J., "Palaeontology and palaeobiogeography of catfishes", in ARRATIA G., KAPOOR B.G., CHARDON M. *et al.* (eds), *Catfishes*, Science Publishers, Enfield, 2003.

[GIB 99] GIBERT J.M.D., BUATOIS L.A., FREGENAL-MARTINEZ M.A. *et al.*, "The fish trace fossil *Undichna* from the Cretaceous of Spain", *Palaeontology*, vol. 42, pp. 409–427, 1999.

[GIB 13a] GIBSON S.Z., "Biodiversity and evolutionary history of *Lophionotus* (Neopterygii: Semionotiformes) from the Western United States", *Copeia*, vol. 2013, pp. 582–603, 2013.

[GIB 13b] GIBSON S.Z., "A new hump-backed Ginglymodian Fish (Neopterygii, Semionotiformes) from the Upper Triassic Chinle Formation of Southeastern Utah", *Journal of Vertebrate Paleontology*, vol. 33, pp. 1037–1050, 2013.

[GIB 16] GIBSON S.Z., "Redescription and phylogenetic placement of †*Hemicalypterus weiri* Schaeffer, 1967 (Actinopterygii, Neopterygii) from the Triassic Chinle Formation, Southwestern United States: new insights into morphology, ecological niche, and phylogeny", *PLoS One*, vol. 11, p. e0163657, 2016.

[GIE 13] GIERL C., REICHENBACHER B., GAUDANT J. *et al.*, "An extraordinary gobioid fish fossil from southern France", *PLoS One*, vol. 8, no. 5, p. e64117, 2013.

[GOB 06] GOBETZ K., LUCAS S.G., LERNER A.J., "Lungfish burrows of varying morphology from the Upper Triassic Redonda Formation, Chinle Group, eastern New Mexico", *New Mexico Museum of Natural History and Science Bulletin*, vol. 37, pp. 140–146, 2006.

[GOT 98] GOTTFRIED M.D., KRAUSE D.W., "First record of gars (Lepisosteidae, Actinopterygii) on Madagascar: Late Cretaceous remains from the Mahajanga basin", *Journal of Vertebrate Paleontology*, vol. 18, no. 2, pp. 275–279, 1998.

[GOT 04] GOTTFRIED M.D., ROGERS R.R., CURRY ROGERS K., "First record of Late Cretaceous coelacanths from Madagascar", in ARRATIA G., WILSON M.V.H., CLOUTIER R. (eds), *Recent Advances in the Origin and Early Radiation of Vertebrates*, Dr. Friedrich Pfeil, Munich, 2004.

[GOT 08] GOTTFRIED M.D., OSTROWSKI S., "Fossil fishing one piece at a time, with a catfish example from the Late Creatceous of Madagascar" *Journal of Vertebrate Paleontology*, vol. 28, pp. 85A–86A, 2008.

[GOT 09] GOTTFRIED M.D., STEVENS N.J., ROBERTS E.M. *et al.*, "A new Cretaceous lungfish (Dipnoi: Ceratodontidae) from the Rukwa Rift Basin, Tanzania", *African Natural History*, vol. 5, pp. 31–36, 2009.

[GOU 11] GOUJET D., "'Lungs' in Placoderms, a persistent palaeobiological myth related to environmental preconceived interpretations", *Comptes Rendus Palevol*, vol. 10, pp. 323–329, 2011.

[GRA 91] GRANDE L., BEMIS E., "Osteology and phylogenetic relationships of fossil and Recent paddlefishes (Polyodontidae) with comments on the interrelationships of Acipenseriformes", *Journal of Vertebrate Paleontology*, vol. 11, pp. 1–121, 1991.

[GRA 97] GRAHAM J.B., *Air-Breathing Fishes: Evolution, Diversity, and Adaptation*, Academic Press, San Diego, 1997.

[GRA 98] GRANDE L., BEMIS W.E., "A comprehensive phylogenetic study of Amiid fishes (Amiidae) based on comparative skeletal anatomy. An empirical search for interconnected patterns of natural history", *Supplement Journal of Vertebrate Palaeontology, Memoir 4*, vol. 18, pp. 1–690, 1998.

[GRA 99] GRANDE L., GRANDE T., "A new species of †*Notogoneus* (Teleostei: Gonorynchidae) from the Upper Cretaceous Two Medicine Formation of Montana, and the poor Cretaceous record of freshwater fishes from North America", *Journal of Vertebrate Paleontology*, vol. 19, pp. 612–622, 1999.

[GRA 82] GRANDE L., "A revision of the fossil genus *Diplomystus*: with comments on the interrelationships of clupeomorph fishes", *American Museum Novitates*, vol. 2728, pp. 1–38, 1982.

[GRA 84] GRANDE L., "Paleontology of the Green River Formation with a review of the fish fauna", *Bulletin of the Geological Survey of Wyoming*, vol. 63, pp. 1–333, 1984.

[GRA 85] GRANDE L., "Recent and Fossil Clupeomorph Fishes with material for revision of the subgroups of clupeoids", *Bulletin of the American Mueum of Natural History*, vol. 181, pp. 231–272, 1985.

[GRA 99] GRANDE T., POYATO-ARIZA F.J., "Phylogenetic relationships of fossil and Recent gonorynchiform fishes (Teleostei: Ostariopysi)", *Zoological Journal of the Linnean Society*, pp. 197–238, 1999.

[GRA 02] GRANDE L., JIN F., YABUMOTO Y. *et al.*, "*Protopsephurus liui*, a well-preserved primitive paddlefish (Acipenseriformes: Polyodontidae) from the Lower Cretaceous of China", *Journal of Vertebrate Paleontology*, vol. 22, pp. 209–237, 2002.

[GRA 04] GRANDE T., DE PINNA M., "The evolution of the Weberian apparatus: a phylogenetic perspective", *Mesozoic Fishes*, vol. 3, pp. 429–448, 2004.

[GRA 06] GRANDE L., HILTON E.J., "An exquisitely preserved skeleton representing a primitive sturgeon from the Upper Cretaceous Judith River Formation of Montana (Acipenseriformes: Acipenseridae: n. gen. and sp)", *Journal of Paleontology*, vol. 80, pp. 1–39, 2006.

[GRA 10] GRANDE L., An empirical synthetic pattern study of gars (Lepisosteiformes) and closely related species, based mostly on skeletal anatomy. The resurrection of holostei. *American Society of Ichthyologists and Herpetologists*, vol. 10, no. 2a, pp. 1–871, 2010.

[GRA 12] GRANDSTAFF B.S., SMITH J.B., LAMANNA M.C. *et al.*, "*Bawitius*, gen. nov., a giant polypterid (Osteichthyes, Actinopterygii) from the Upper Cretaceous Bahariya Formation of Egypt", *Journal of Vertebrate Paleontology*, vol. 32, pp. 17–26, 2012.

[GRE 60] GREENWOOD P.H., *Fossil Denticipitid Fishes from East Africa*, British Museum (Natural History), London, 1960.

[GRE 72] GREENWOOD P.H., "Fish fossils from the Late Miocene of Tunisia", *Notes du Service géologique de Tunisie*, vol. 37, pp. 41–72, 1972.

[GRE 73] GREENWOOD P.H., "Interrelationships of osteoglossomorphs", in GREENWOOD P.H., MILES R.S.P.C. (eds), *Interrelationships of Fishes*, Academic Press, San Diego, 1973.

[GRI 99] GRIGORESCU D., VENCZEL M., CSIKI Z. *et al.*, "New latest Cretaceous microvertebrate fossil assemblages from the Haţeg Basin (Romania)", *Geologie en Mijnbouw*, vol. 78, pp. 301–314, 1999.

[GRO 12] GROSBERG R.K., VERMEIJ G.J., WAINWRIGHT P.C., "Biodiversity in water and on land", *Current Biology*, vol. 22, pp. R900–R903, 2012.

[GUE 14] GUERIAU P., MOCUTA C., DUTHEIL D.B. *et al.*, "Trace elemental imaging of rare earth elements discriminates tissues at microscale in flat fossils", *PLoS One*, vol. 9, p. e86946, 2014.

[GUI 15a] GUINOT G., CAVIN L., "Contrasting 'fish' diversity dynamics between marine and freshwater environments", *Current Biology*, vol. 25, no 17, 2314–2318, 2015.

[GUI 15b] GUINOT G., CAVIN L. "'Fish' (Actinopterygii and Elasmobranchii) diversification patterns through deep time", *Biological Reviews*, 2015.

[HIL 03] HILTON E.J., "Comparative osteology and phylogenetic systematics of fossil and living bony-tongue fishes (Actinopterygii, Teleostei, Osteoglossomorpha)", *Zoological Journal of the Linnean Society*, vol. 137, pp. 1–100, 2003.

[HIL 04] HILTON E.J., GRANDE L., BEMIS W.E., "Morphology of †*Coccolepis bucklandi* Agassiz, 1843 (Actinopterygii, †Coccolepidae) from the Solnhofen Lithographic Limestone deposits (Upper Jurassic, Germany)", in ARRATIA G., TINTORI A. (eds), *Mesozoic Fishes 3 – Systematics, Paleoenvironments and Biodiverity*, Dr. Friedrich Pfeil, Munich, 2004.

[HIL 08] HILTON E.J., GRANDE L., "Fossil mooneyes (Teleostei: Hiodontiformes, Hiodontidae) from the Eocene of western North America, with a reassessment of their taxonomy", *Geological Society, London, Special Publications*, vol. 295, pp. 221–251, 2008.

[HOR 05] HORSTKOTTE J., STRECKER U., "Trophic differentiation in the phylogenetically young *Cyprinodon* species flock (Cyprinodontidae, Teleostei) from Laguna Chichancanab (Mexico)", *Biological Journal of the Linnean Society*, vol. 85, no. 1, pp. 125–134, 2005.

[IBR 14] IBRAHIM N., SERENO P., DAL SASSO C. *et al.*, "Semiaquatic adaptations in a giant predatory dinosaur", *Science*, vol. 345, no. 6204, pp. 1613–1616, 2014.

[IMO 13] IMOTO J.M., SAITOH K., SASAKI T. *et al.*, "Phylogeny and biogeography of highly diverged freshwater fish species (Leuciscinae, Cyprinidae, Teleostei) inferred from mitochondrial genome analysis", *Gene*, vol. 514, pp. 112–124, 2013.

[INO 09] INOUE J.G., KUMAZAWA Y., MIYA M. *et al.*, The historical biogeography of the freshwater knifefishes using mitogenomic approaches: a Mesozoic origin of the Asian notopterids (Actinopterygii: Osteoglossomorpha)", *Molecular Phylogenetics and Evolution*, vol. 51, no. 3, pp. 486–499, 2009.

[JAI 83a] JAIN S.L., "A review of the genus *Lepidotes* (Actinopterygii: Semionotiformes) with special references to the species from Kota Formation (Lower Jurassic), India", *Journal of the Palaeontological Society of India*, vol. 28, pp. 7–42, 1983.

[JAI 83b] JAIN S.L., SAHNI A., "Some Cretaceous vertebrates from Central India and their palaeogeographic implications", *Indian Association for Palynostratigraphy, Symposium Abstract Volume, Lucknow*, pp. 1–16, 1983.

[JAN 07] JANVIER P., DESBIENS S., WILLETT J.A., "New evidence for the controversial "lungs" of the Late Devonian antiarch *Bothriolepis canadensis* (Whiteaves, 1880) (Placodermi: Antiarcha)", *Journal of Vertebrate Paleontology*, vol. 27, pp. 709–710, 2007.

[JIN 06] JIN F., "An overview of Triassic fishes from China", *Vert PalAsiat*, vol. 44, pp. 28–42, 2006.

[JUB 75] JUBB R.A., GARDINER B.G., "A preliminary catalogue of identifiable fossil fish material from southern Africa", *Annals of the South African Museum*, vol. 67, pp. 1-59, 1975.

[KIR 98] KIRKLAND J.I., "Morrison Fishes", *Modern Geology*, vol. 22, pp. 503–533, 1998.

[KEM 92] KEMP A., "New neoceratodont cranial remains from the Late Oligocene-Middle Miocene of northern Australia with comments on generic characters for Cenozoic lungfish", *Journal of Vertebrate Paleontology*, vol. 12, pp. 284–293, 1992.

[KEM 94] KEMP A., "Australian Triassic lungfish skulls", *Journal of Paleontology*, vol. 68, no. 3, pp. 647–654, 1994.

[KEM 97a] KEMP A., "Four species of *Metaceratodus* (Osteichthyes: Dipnoi, Family Ceratodontidae) from Australian Mesozoic and Cenozoic deposits", *Journal of Vertebrate Paleontology*, vol. 17, no. 1, pp. 26–33, 1997.

[KEM 97b] KEMP A., "A revision of Australian Mesozoic and Cenozoic lungfish of the family Neoceratodontidae (Osteichthyes: Dipnoi), with a description of four new species", *Journal of Paleontology*, vol. 71, no. 4, pp. 713–733, 1997.

[KEM 98] KEMP A., "Skull structure in post-paleozoic lungfish", *Journal of Vertebrate Paleontology*, vol. 18, pp. 43–63, 1998.

[KEM 05] KEMP A., "New insights into ancient environments using dental characters in Australian Cenozoic lungfish", *Alcheringa*, vol. 29, pp. 123–149, Kemps 17, 2005.

[KHA 10] KHALLOUFI B., OUARHACHE D., LELIÈVRE H., "New paleontological and geological data about Jbel Tselfat (Late Cretaceous of Morocco)", *Historical Biology*, vol. 22, pp. 57–70, 2010.

[KIL 31] KILIAN C., "Des principaux complexes continentaux du Sahara", *Comptes Rendus sommaire de la Société Géologique de France*, vol. 9, pp. 109–111, 1931.

[KIM 14] KIM H.M., CHANG M.-M., WU F. et al., "A new ichthyodectiform (Pisces, Teleostei) from the Lower Cretaceous of South Korea and its paleobiogeographic implication", *Cretaceous Research*, vol. 47, pp. 117–130, 2014.

[KOC 09] KOCSIS L., ÖSI A., VENNEMANN T. et al., "Geochemical study of vertebrate fossils from the Upper Cretaceous (Santonian) Csehbánya Formation (Hungary): evidence for a freshwater habitat of mosasaurs and pycnodont fish", *Palaeogeography Palaeoclimatology Palaeoecology*, vol. 280, nos. 3–4, pp. 532–542, 2009.

[KOG 09] KOGAN I., SCHÖNBERGER K., FISCHER J. *et al.*, "A nearly complete skeleton of *Saurichthys orientalis* (Pisces, Actinopterygii) from the Madygen Formation (Middle to Late Triassic, Kyrgyzstan, Central Asia)–preliminary results", *Freiberger Forschungshefte C*, vol. 532, pp. 139–152, 2009.

[KON 03] KONTULA T., KIRILCHIK S.V., VÄINÖLÄ R., "Endemic diversification of the monophyletic cottoid fish species flock in Lake Baikal explored with mtDNA sequencing", *Molecular Phylogenetics and Evolution*, vol. 27, pp. 143–155, 2003.

[KOT 06] KOTTELAT M., BRITZ R., HUI T.H. *et al.*, "*Paedocypris*, a new genus of Southeast Asian cyprinid fish with a remarkable sexual dimorphism, comprises the world's smallest vertebrate", *Proceedings of the Royal Society of London B: Biological Sciences*, vol. 273, no. 1589, pp. 895–899, 2006.

[KRA 06] KRAUSE D.W., O'CONNOR P.M., ROGERS K.C. *et al.*, "Late Cretaceous terrestrial vertebrates from Madagascar: Implications for latin American biogeography", *Annals of the Missouri Botanical Garden*, vol. 93, no. 2, pp. 178–208, 2006.

[KRI 05] KRIWET J., "A comprehensive study of the skull and dentition of pycnodont fishes", *Zitteliana*, vol. A45, pp. 135–188, 2005.

[KUM 00] KUMAZAWA Y., NISHIDA M., "Molecular phylogeny of Osteoglossoids: a new model for Gondwanian origin and plate tectonic transportation of the Asian Arowana", *Molecular Biology and Evolution*, vol. 17, pp. 1869–1878, 2000.

[KUM 05] KUMAR K., RANA R.S., PALIWAL B.S., "Osteoglossid and lepisosteid fish remains from the Paleocene Palana Formation, Rajasthan, India", *Palaeontology*, vol. 48, pp. 1187–1209, 2005.

[LÄN 13] LÄNG E., BOUDAD L., MAIO L. *et al.*, "Unbalanced food web in a Late Cretaceous dinosaur assemblage", *Palaeogeography, Palaeoclimatology, Palaeoecology*, vol. 381–382, pp. 26–32, 2013.

[LAU 16] LAUMANN K.M., Sturgeon (Acipenseridae) Phylogeny, Biogeography, & Ontogeny, The College of William & Mary, Williamburg, 2016.

[LAV 54] LAVOCAT R., *Reconnaissance géologique dans les Hammadas des confins algéro-marocains du sud*, Editions du service géologique du Maroc, Rabat 1954.

[LAV 04] LAVOUÉ S., SULLIVAN J.P., "Simultaneous analysis of five molecular markers provides a well-supported phylogenetic hypothesis for the living bony-tongue fishes (Osteoglossomorpha: Teleostei)", *Molecular Phylogenetics and Evolution*, vol. 33, pp. 171–185, 2004.

[LAV 12] LAVOUÉ S., MIYA M., ARNEGARD M.E. *et al.*, "Comparable ages for the independent origins of electrogenesis in African and South American weakly electric fishes", *PLoS One*, vol. 7, p. e36287, 2012.

[LAV 13] LAVOUÉ S., MIYA M., MUSIKASINTHORN P. *et al.*, "Mitogenomic evidence for an indo-west pacific origin of the clupeoidei (Teleostei: Clupeiformes)", *PLoS One*, vol. 8, p. e56485, 2013.

[LAV 15] LAVOUÉ S., "Testing a time hypothesis in the biogeography of the arowana genus *Scleropages* (Osteoglossidae)", *Journal of Biogeography*, vol. 42, pp. 2427–2439, 2015.

[LAV 16] LAVOUÉ S., "Was Gondwanan breakup the cause of the intercontinental distribution of Osteoglossiformes? A time-calibrated phylogenetic test combining molecular, morphological, and paleontological evidence", *Molecular Phylogenetics and Evolution*, vol. 99, pp. 34–43, 2016.

[LEA 04] LEAL M.E.D.C., BRITO I.M., "The ichthyodectiform *Cladocyclus gardneri* (Actinopterygii: Teleostei) from the Crato and Santana Formations, Lower Cretaceous of Araripe Basin, North-Eastern Brazil", *Annales de Paléontologie*, vol. 90, pp. 103–113, 2004.

[LEE 07] LEE D.E., MCDOWALL R.M., LINDQVIST J.K., "*Galaxias* fossils from Miocene lake deposits, Otago, New Zealand: the earliest records of the Southern Hemisphere family Galaxiidae (Teleostei)", *Journal of the Royal Society of New Zealand*, vol. 37, no. 3, pp. 109–130, 2007.

[LEL 12] LE LOEUFF J., LÄNG E., CAVIN L. *et al.*, "Between Tendaguru and Bahariya: on the age of the Early Cretaceous Dinosaurs sites from the Continental Intercalaire and other African formations", *Journal of Stratigraphy*, vol. 36, pp. 1–18, 2012.

[LI 96] LI G.Q., "A new species of Late Cretaceous Osteoglossid (Teleostei) from the Oldman Formation of Alberta, Canada, and its phylogenetic relationships", in ARRATIA G., VIOHL G. (eds), *Mesozoic Fishes 1 – Systematics and Paleoecology*, Dr. Friedrich Pfeil, Munich, 1996.

[LI 97] LI G.-Q., GRANDE L., WILSON M.V.H., "The species of †*Phareodus* (Teleostei: Osteoglossidae) from the Eocene of North America and their phylogenetic relationships", *Journal of Vertebrate Paleontology*, vol. 17, no. 3, pp. 487–505, 1997.

[LI 99] LI G.Q., WILSON M.V.H., "Early divergence of Hiodontiformes sensu stricto in East Asia and phylogeny of some Late Mesozoic teleosts from China", in ARRATIA G., SCHULTZE H.-P. (eds), *Mesozoic Fishes 2 – Systematics and Fossil Record*, Dr. Friedrich Pfeil, Munich, 1999.

[LI 06] LI X., MUSIKASINTHORN P., KUMAZAWA Y., "Molecular phylogenetic analyses of snakeheads (Perciformes: Channidae) using mitochondrial DNA sequences", *Ichthyological Research*, vol. 53, pp. 148–159, 2006.

[LI 10] LI J., XIA R., MCDOWALL R. *et al.*, "Phylogenetic position of the enigmatic *Lepidogalaxias salamandroides* with comment on the orders of lower euteleostean fishes", *Molecular Phylogenetics and Evolution*, vol. 57, pp. 932–936, 2010.

[LIU 09] LIU J., CHANG M.-M., "A new Eocene catostomid (Teleostei: Cypriniformes) from northeastern China and early divergence of Catostomidae", *Science in China Series D: Earth Sciences*, vol. 52, pp. 189–202, 2009.

[LOM 13] LOMBARDO C., "A new basal actinopterygian fish from the Late Ladinian of Monte San Giorgio (Canton Ticino, Switzerland)", *Swiss Journal of Geosciences*, vol. 106, pp. 219–230, 2013.

[LON 84] LONGBOTTOM A.E., "New tertiary pycnodants from the Tilemsi valley, Republic of Mali", *Bulletin of The British Museum (Natural History) Geology*, vol. 38, pp. 1–26,1984.

[LON 10] LONGBOTTOM A., "A new species of the catfish *Nigerium* from the Palaeogene of the Tilemsi valley, Republic of Mali", *Palaeontology*, vol. 53, pp. 571–594, 2010.

[LON 17] LONGRICH N.R., "A Stem Lepidosireniform Lungfish (Sarcopterygia: Dipnoi) from the Upper Eocene of Libya, North Africa and implications for Cenozoic lungfish evolution", *Gondwana Research*, vol. 42, pp. 140–150, 2017.

[LÓP 04] LÓPEZ-ARBARELLO A., "The record of Mesozoic fishes from Gondwana (excluding India and Madagascar)", in ARRATIA, G., TINTORI, A. (eds), *Mesozoic Fishes 3 – Systematics, Paleoenvironments and Biodiversity*, Dr. Friedrich Pfeil, Munich, 2004.

[LÓP 06] LÓPEZ-ARBARELLO A., ROGERS R., PUERTA P., "Freshwater actinopterygians of the Los Rastros Formation (Triassic), Bermejo Basin, Argentina", *Fossil Record*, vol. 9, pp. 238–258, 2006.

[LÓP 07] LÓPEZ-ARBARELLO A., CODORNIÚ L., "Semionotids (Neopterygii, Semionotiformes) from the Lower Cretaceous Lagarcito Formation, San Luis Province, Argentina", *Journal of Vertebrate Paleontology*, vol. 27, no. 4, pp. 811–826, 2007.

[LÓP 08a] LÓPEZ-ARBARELLO A., RAUHUT O.W.M., MOSER K., "Jurassic fishes of Gondwana", *Revista de la Asociacion Geologica Argentina*, vol. 63, pp. 586–612, 2008.

[LÓP 08b] LÓPEZ-ARBARELLO A., ZAVATTIERI A.M., "Systematic revision of *Pseudobeaconia* Bordas, 1944, and Mendocinichthys Whitley, 1953 (Actinopterygii:'Perleidiformes') from the Triassic of Argentina", *Palaeontology*, vol. 51, pp. 1025–1052, 2008.

[LÓP 10] LÓPEZ-ARBARELLO A., RAUHUT O.W., CERDENO E., "The Triassic fish faunas of the Cuyana Basin, western Argentina", *Palaeontology*, vol. 53, pp. 249–276, 2010.

[LÓP 12] LÓPEZ-ARBARELLO A., "Phylogenetic interrelationships of ginglymodian fishes (Actinopterygii: Neopterygii)", *PLoS One*, vol. 7, p. e39370, 2012.

[LÓP 13] LÓPEZ-ARBARELLO A., SFERCO E., RAUHUT O.W., "A new genus of coccolepidid fishes (Actinopterygii, Chondrostei) from the continental Jurassic of Patagonia", *Palaeontologia Electronica*, vol. 16, pp. 1–23, 2013.

[LOV 06] LOVEJOY N.R., ALBERT J.S., CRAMPTON W.G., "Miocene marine incursions and marine/freshwater transitions: evidence from Neotropical fishes", *Journal of South American Earth Sciences*, vol. 21, no. 1, pp. 5–13, 2006.

[LUN 93] LUNDBERG J.G., "African-South American freshwater fish clades and continental drift: problems with a paradigm", in GOLDBLATT P. (ed.), *Biological Relationships Between Africa and South America*, Yale University Press, New Haven, 1993.

[LUN 98a] LUNDBERG J.G., "The temporal context for the diversification of neotropical fishes", in MALABARBA L.R., REIS R.E., VARI R.P. *et al.* (eds), *Phylogeny and Classification of Neotropical Fishes*, EDIPUCRS, Porto Alegre, 1998.

[LUN 98b] LUNDBERG J.G., MARSHALL L.G., GUERRERO J. *et al.*, "The stage for Neotropical fish diversification: a history of tropical South American rivers", in MALABARBA L.R., REIS R.E., VARI R.P. *et al.* (eds), *Phylogeny and Classification of Neotropical Fishes*, EDIPUCRS, Porto Alegre, 1998.

[MAC 63] MACARTHUR R.H., WILSON E.O., "An equilibrium theory of insular zoogeography", *Evolution*, vol. 17, pp. 373–387, 1963.

[MAC 67] MACARTHUR R.H., WILSON E.O., *The Theory of Island Biogeography*, Princeton University Press, 1967.

[MAC 14] MACQUEEN D.J., JOHNSTON I.A., "A well-constrained estimate for the timing of the salmonid whole genome duplication reveals major decoupling from species diversification", *Proceedings of the Royal Society B*, vol. 281, no. 1778, pp. 1–8, 2014.

[MAI 03] MAISCH M.W., MATZKE A.T., PFRETZSCHNER H.U. *et al.*, "Fossil vertebrates from the Middle and Upper Jurassic of the Southern Junggar Basin (NW China)-results of the Sino-German Expeditions 1999-2000", N*eues Jahrbuch für Geologie und Paläontologie Monatshefte*, vol. 5, pp. 297–313, 2003.

[MAI 86] MAISEY J.G., "Coelacanths from the Lower Cretaceous of Brazil", *American Museum Novitates*, vol. 2866, pp. 1–30, 1986.

[MAI 91] MAISEY J.G., "*Cladocyclus* Agassiz, 1841", in MAISEY J.G. (ed.), *Santana Fossils. An Illustrated Atlas*, TFH Publication, New York, 1991.

[MAI 94] MAISEY J.G., "Predator–prey relationships and trophic level reconstruction in a fossil fish community", *Environmental Biology of Fishes*, vol. 40, pp. 1–22, 1994.

[MAI 00] MAISEY J.G., "Continental break up and the distribution of fishes of Western Gondwana during the Early Cretaceous", *Cretaceous Research*, vol. 21, no. 2-3, pp. 281–314, 2000.

[MAI 11] MAISEY J., "Northeastern Brazil: out of Africa", *Paleontologia: Cenários de vida*, vol. 4, pp. 545–559, 2011.

[MAI 16] MAISEY J.G., "Mr Mawson's fossils, *Geological Society, London, Special Publications*, vol. 430, pp. 219–233, 2016.

[MAL 06] MALABARBA M.C., ZULETA O., PAPA C.D., "*Proterocara argentina*, a new fossil cichlid from the Lumbrera Formation, Eocene of Argentina", *Journal of Vertebrate Paleontology*, vol. 26, pp. 267–275, 2006.

[MAL 10a] MALABARBA L., MALABARBA M., "Biogeography of Characiformes: an evaluation of the available information of fossil and extant taxa", in NELSON J.S., SCHULTZE H.-P., WILSON M.V.H. (eds), *Origin and Phylogenetic Interrelationships of Teleosts*, Dr. Friedrich Pfeil, Munich, 2010.

[MAL 10b] MALABARBA M.C., MALABARBA L.R., PAPA C.D., "*Gymnogeophagus eocenicus*, n. sp. (Perciformes: Cichlidae), an Eocene cichlid from the Lumbrera Formation in Argentina", *Journal of Vertebrate Paleontology*, vol. 30, pp. 341–350, 2010.

[MAR 80] MARTIN M., "*Mauritanichthys rugosus* n. gen. et n. sp., Redfieldiidae (Actinopterygi, chondrostei), du Trias superieur continental marocain", *Geobios*, vol. 13, pp. 437–440, 1980.

[MAR 81] MARTIN M., "Les dipneustes et Actinistiens du Trias supérieur continental marocain", *Stuttgarter Beiträge zur Naturkunde, Serie B (Geologie und Paläontologie)*, vol. 69, pp. 1–29, 1981.

[MAR 82] MARTIN M., "Nouvelles données sur la phylogénie et la systématique des dipnoi postpaléozoïques", *Comptes Rendus de l'Académie des Sciences, Paris, série II*, vol. 294, pp. 611–614, 1982.

[MAR 95] MARTIN M., "Nouveaux lepidosirenidés (Dipnoi) du Tertiaire africain", *Geobios*, vol. 28, pp. 275–280, 1995.

[MAR 97] MARTENS K., "Speciation in ancient lakes", *Trends in Ecology & Evolution*, vol. 12, pp. 177–182, 1997.

[MAR 15] MARTINELLI A., TEIXEIRA V., "The Late Cretaceous vertebrate record from the Bauru Group in the Triângulo Mineiro, southeastern Brazil", *Boletín Geológico y Minero*, vol. 126, pp. 129–158, 2015.

[MAW 07] MAWSON J., WOODWARD A.S., "Cretaceous formation of Bahia and its vertebrate fossils", *Quarterly Journal of the Geological Society*, vol. 63, pp. 128–139, 1907.

[MAY 94] MAY R.M., GODFREY J., "Biological diversity: differences between land and sea [and discussion]", *Philosophical Transactions of the Royal Society of London B: Biological Sciences*, vol. 343, pp. 105–111, 1994.

[MAY 09] MAYDEN R.L., CHEN W.-J., BART H.L. *et al.*, "Reconstructing the phylogenetic relationships of the Earth's most diverse clade of freshwater fishes – order Cypriniformes (Actinopterygii: Ostariophysi): a case study using multiple nuclear loci and the mitochondrial genome", *Molecular and Phylogenetic Evolution*, vol. 51, pp. 500–514, 2009.

[MAY 10] MAYDEN R.L., CHEN W.-J., "The world's smallest vertebrate species of the genus *Paedocypris*: a new family of freshwater fishes and the sister group to the world's most diverse clade of freshwater fishes (Teleostei: Cypriniformes)", *Molecular Phylogenetics and Evolution*, vol. 57, pp. 152–175, 2010.

[MAY 15] MAYRINCK D., BRITO P., OTERO O., "Review of the osteology of the fossil fish formerly attributed to the genus †*Chanoides* and implications for the definition of otophysan bony characters", *Journal of Systematic Palaeontology*, vol. 13, pp. 397–420, 2015.

[MCC 87] MCCUNE A.R., "Toward the phylogeny of a fossil species Flock: Semionotid fishes from a Lake Deposit in the Early Jurassic Towaco Formation, Newark Basin", *Peabody Museum of Natural History, Yale University*, vol. 43, pp. 1–108, 1987.

[MCC 96] MCCUNE A.R., "Biogeographic and stratigraphic evidence for rapid speciation in semionotid fishes", *Paleobiology*, vol. 22, pp. 34–48, 1996.

[MCD 97] MCDOWALL R.M., POLE M., "A large galaxiid fossil (Teleostei) from the Miocene of Central Otago, New Zealand", *Journal of the Royal Society of NZ*, vol. 27, pp. 193–198, 1997.

[MCD 02] MCDOWALL R.M., "Accumulating evidence for a dispersal biogeography of southern cool temperate freshwater fishes", *Journal of Biogeography*, vol. 29, pp. 207–219, 2002.

[MCD 06] MCDOWALL R.M., KENNEDY E.M., LINDQVIST J.K. *et al.*, "Probable Gobiomorphus fossils from the Miocene and Pleistocene of New Zealand (Teleostei: Eleotridae)", *Journal of the Royal Society of New Zealand*, vol. 36, no. 3, pp. 97–109, 2006.

[MCD 10] MCDOWALL R.M., *New Zealand Freshwater Fishes: A Historical and Ecological Biogeography,* vol. 32, Springer, Heidelberg, 2010.

[MEU 96] MEUNIER F.J., GAYET M., "A new polypteriform from the Late Cretaceous and the Middle Paleocene of South America", in ARRATIA G., VIOHL G. (eds), *Mesozoic Fishes 1 – Systematics and Paleoecology*, Dr. Friedrich Pfeil, Munich, 1996.

[MEU 98] MEUNIER F.J., GAYET M., "Rectification of the nomenclature of the genus name *Pollia* Meunier & Gayet, 1996 (Osteichthyes, Cladistia, Polypteriformes) in *Latinopollia* nom. nov.", *Cybium*, vol. 22, p. 192, 1998.

[MEU 13] MEUNIER F.J., DUTHEIL D.B., BRITO P.M., "Histological study of the median lingual dental plate of the Cretaceous fish †*Palaeonotopterus greenwoodi* (Teleostei: Osteoglossomorpha) from the Kem-Kem beds, Morocco", *Cybium*, vol. 37, pp. 121–125, 2013.

[MEU 16] MEUNIER F.J., EUSTACHE R.-P., DUTHEIL D. *et al.*, "Histology of ganoid scales from the early Late Cretaceous of the Kem Kern beds, SE Morocco: systematic and evolutionary implications", *Cybium*, vol. 40, pp. 121–132, 2016.

[MEY 87] MEYER A., Phenotypic plasticity and heterochrony in *Cichlasoma managuense* (Pisces, Chichlidae) and their implications for speciation in Cichlid fishes, *Evolution*, vol. 41, no. 6, pp. 1357–1369, 1987.

[MIC 01] MICKLICH N., KLAPPERT G., "*Masillosteus kelleri*, a new gar (Actinopterygii, Lepisosteidae) from the middle Eocene of Grube Messel (Hessen, Germany)", *Kaupia*, vol. 11, pp. 73–81, 2001.

[MIC 12] MICKLICH N., "Peculiarities of the Messel fish fauna and their palaeoecological implications: a case study", *Palaeobiodiversity and Palaeoenvironments*, vol. 92, pp. 585–629, 2012.

[MIL 06a] MILNER A.R., KIRKLAND J.I., "Preliminary review of the early Jurassic (Hettangian) freshwater Lake Dixie fish fauna in the Whitmore Point Member, Moenave Formation in southwest Utah. Southwest Utah", *New Mexico Museum of Natural History and Science Bulletin*, vol. 37, pp. 510–521, 2006.

[MIL 06b] MILNER A.R., KIRKLAND J.I., BIRTHISEL. T.A., "The geographic distribution and biostratigraphy of the Late Triassic-Early Jurassic freshwater fish faunas of the southwestern United States", *New Mexico Museum of Natural History and Science Bulletin*, vol. 37, pp. 522–529, 2006.

[MKH 11] MKHITARYAN T.G., AVERIANOV A.O., "New material and phylogenetic position of *Aidachar paludalis* Nesov, 1981 (Actinopterygii, Ichthyodectiformes) from the Late Cretaceous of Uzbekistan", *Proceedings of the Zoological Institute RAS*, vol. 315, pp. 181–192, 2011.

[MO 16] MO J., BUFFETAUT E., TONG H. *et al.*, "Early Cretaceous vertebrates from the Xinlong Formation of Guangxi (southern China): a review", *Geological Magazine*, vol. 153, no. 1, pp. 143–159, 2016.

[MOG 09] MOGUTCHEVA N., KRUGOVYKH V., "New data on the stratigraphic chart for Triassic deposits in the Tunguska syneclise and Kuznetsk Basin", *Stratigraphy and Geological Correlation*, vol. 17, pp. 510–518, 2009.

[MOH 96] MOHABEY D.M., UDHOJI S.G., "*Pycnodus lametae* (Pycnodontidae), a holostean fish from freshwater Upper Cretaceous Lameta Formation of Maharashtra", *Geological Society of India*, vol. 47, no. 5, pp. 593–598, 1996.

[MOR 04] MORLO M., SCHAAL S., MAYR G. *et al.*, "An annotated taxonomic list of the Middle Eocene (MP 11) Vertebrata of Messel", *Courier Forschungsinstitut Senckenberg* , vol. 252, pp. 95–108, 2004.

[MUR 97] MURPHY W.J., COLLIER G.E., "A molecular phylogeny for aplocheiloid fishes (Atherinomorpha, Cyprinodontiformes): the role of vicariance and the origins of annualism", *Molecular Biology and Evolution*, vol. 14, no. 8, pp. 790–799, 1997.

[MUR 00] MURRAY A.M., "The palaeozoic, mesozoic and early Cenozoic fishes of Africa", *Fish and Fisheries*, vol. 1, pp. 111–145, 2000.

[MUR 01a] MURRAY A.M., "Eocene cichlid fishes from Tanzania, East Africa", *Journal of Vertebrate Paleontology*, vol. 20, pp. 651–664, 2001.

[MUR 01b] MURRAY A.M., "The fossil record and biogeography of the Cichlidae (Actinopterygii: Labroidei)", *Biological Journal of the Linnean Society*, vol. 74, pp. 517–532, 2001.

[MUR 01c] MURRAY A.M., "The oldest fossil cichlids (Teleostei: Perciformes): indication of a 45 million-year-old species flock", *Proceedings of Royal Society of London Series B*, vol. 268, pp. 679–684, 2001.

[MUR 02] MURRAY A.M., "Lower pharyngeal jaw of a cichlid fish (Actinopterygii; Labroidei) from an early Oligocene site in the Fayum, Egypt", *Journal of Vertebrate Paleontology*, vol. 22, pp. 453–455, 2002.

[MUR 03a] MURRAY A.M., "A new characiform fish (Teleostei: Ostariophysi) from the Eocene of Tanzania", *Canadian Journal of Earth Sciences*, vol. 40, pp. 473–481, 2003.

[MUR 03b] MURRAY A.M., "A new Eocene citharinoid fish (Ostariophysi: Characiformes) from Tanzania", *Journal of Vertebrate Paleontology*, vol. 23, pp. 501–507, 2003.

[MUR 03c] MURRAY A.M., BUDNEY, L.A., "A new species of catfish (Claroteidae, *Chrysichthys*) from an Eocene crater lake in East Africa", *Canadian Journal of Earth Sciences*, vol. 40, pp. 983–993, 2003.

[MUR 04] MURRAY A.M., "Late Eocene and early Oligocene teleost and associated ichthyofauna of the Jebel Qatrani Formation, Fayum, Egypt", *Palaeontology*, vol. 47, pp. 711–724, 2004.

[MUR 05a] MURRAY A.M., SIMONS E.L., ATTIA Y.S., "A new clupeid fish (Clupeomorpha) from the Oligocene of Fayum, Egypt, with notes on some other fossil clupeomorphs", *Journal of Vertebrate Paleontology*, vol. 25, pp. 300–308, 2005.

[MUR 05b] MURRAY A.M., WILSON M.V., "Description of a new Eocene osteoglossid fish and additional information on †*Singida jacksonoides* Greenwood and Patterson, 1967 (Osteoglossomorpha), with an assessment of their phylogenetic relationships", *Zoological Journal of the Linnean Society*, vol. 144, pp. 213–228, 2005.

[MUR 06] MURRAY A.M., "A new channid (Teleostei: Channiformes) from the Eocene and Oligocene of Egypt", *Journal of Paleontology*, vol. 80, no. 6, pp. 1172–1178, 2006.

[MUR 08] MURRAY A.M., THEWISSEN J.G.M., "Eocene actinopterygian fishes from Pakistan, with the description of a new genus and species of channid (channiformes)", *Journal of Vertebrate Paleontology*, vol. 28, pp. 41–52, 2008.

[MUR 10a] MURRAY A.M., COOK T.D., ATTIA Y.S. *et al.*, "A freshwater ichthyofauna from the late Eocene Birket Qarun Formation, Fayum, Egypt", *Journal of Vertebrate Paleontology*, vol. 30, pp. 665–680, 2010.

[MUR 10b] MURRAY A.M., YOU H.-L., PENG C., "A new Cretaceous osteoglossomorph fish from Gansu Province, China", *Journal of Vertebrate Paleontology*, vol. 30, pp. 322–332, 2010.

[MUR 13] MURRAY A.M., WILSON M.V.H., "Two new paraclupeid fishes (Clupeomorpha: Ellimmichthyiformes) from the Upper Cretaceous of Morocco", in ARRATIA G., SCHULTZE H.-P., WILSON M.V.H. (eds), *Mesozoic Fishes 5 – Global Diversity and Evolution*, Dr. Friedrich Pfeil, Munich, 2013.

[MUS 11] MUSCHICK M., BARLUENGA M., SALZBURGER W. *et al.*, "Adaptive phenotypic plasticity in the Midas cichlid fish pharyngeal jaw and its relevance in adaptive radiation", *BMC Evolutionary Biology*, vol. 11, pp. 116, 2011.

[MYE 49] MYERS G.S., "Salt-tolerance of fresh-water fish groups in relation to zoogeographical problems", *Bijdragen tot de Dierkunde*, vol. 28, pp. 315–322, 1949.

[NAK 11] NAKATANI M., MIYA M., MABUCHI K. *et al.*, "Evolutionary history of Otophysi (Teleostei), a major clade of the modern freshwater fishes: Pangaean origin and Mesozoic radiation", *BMC Evolutionary Biology*, vol. 11, pp. 1–25, 2011.

[NEA 03] NEAR T.J., KASSLER T.W., KOPPELMAN J.B. *et al.*, "Speciation in North American black basses, *Micropterus* (Actinopterygii: Centrarchidae)", *Evolution*, vol. 57, pp. 1610–1621, 2003.

[NES 85] NESSOV L.A., KAZNYSHKIN M.N., "A lungfish and turtles from Upper Jurassic of Nothern Fergana, Kirghiz SSR", *Vestnik Zoologii*, vol. 1, pp. 33–39, 1985.

[NEU 05] NEUMAN A., BRINKMAN D., "Fishes of the fluvial beds", in CURRIE P.J., KOPPELHUS E.B. (eds), *Dinosaur Provincial Park: A Spectacular Ancient Ecosystem Revealed*, Indiana University Press, Bloomington, 2005.

[NEW 09] NEWBREY M., MURRAY A., WILSON M. *et al.*, "Seventy-five-million-year-old tropical tetra-like fish from Canada tracks Cretaceous global warming", *Proceedings of the Royal Society of London B: Biological Sciences*, vol. 276, 2009.

[NEW 10] NEWBREY M.G., MURRAY A.M., BRINKMAN D.B. *et al.*, "A new articulated freshwater fish (Clupeomorpha, Ellimmichthyiformes) from the Horseshoe Canyon Formation, Maastrichtian, of Alberta, Canada", *Canadian Journal of Earth Sciences*, vol. 47, pp. 1183–1196, 2010.

[NEW 13] NEWBREY M.G., MURRAY A.M., WILSON M.V.H. *et al.*, "A new species of the paracanthopterygian *Xenyllion* (Sphenocephaliformes) from the Mowry Formation (Cenomanian) of Utah, USA", in ARRATIA G., SCHULTZE H.-P., WILSON M.V.H. (eds), *Mesozoic Fishes 5 – Global Diversity and Evolution*, Dr. Friedrich Pfeil, Munich, 2013.

[NOR 00] NORTHCOTE T.G., "Ecological interactions among an Orestiid (Pisces: Cyprinodontidae) species flock in the littoral zone of Lake Titicaca", *Advances in Ecological Research*, vol. 31, pp. 399–420, 2000.

[OLS 82] OLSEN P.E., MCCUNE A.R., THOMSON K.S., "Correlation of the early Mesozoic Newark Supergroup by vertebrates, principally fishes", *American Journal of Science*, vol. 282, pp. 1–44, 1982.

[ORS 07] ORSULAK M., SIMPSON E., WOLF H.L. *et al.*, "A lungfish burrow in Late Cretaceous upper capping sandstone member of the Wahweap Formation Cockscomb area, grand Staircase-Escalante National Monument, Utah", *Geological Society of America*, vol. 39, no. 5, pp. 43, 2007.

[ØST 06] ØSTBYE K., AMUNDSEN P.A., BERNATCHEZ L. *et al.*, "Parallel evolution of ecomorphological traits in the European whitefish *Coregonus lavaretus* (L.) species complex during postglacial times", *Molecular Ecology*, vol. 15, no. 13, pp. 3983–4001, 2006.

[OST 12] OSTROWSKI S.A., The teleost ichthyofaunal from the Late Cretaceous of Madagascar: systematics, distributions, and implications for Gondwanan biogeography, Dissertation, Michigan State University, 2012.

[OTE 99] OTERO O., GAYET M. "*Weilerichthys fajumensis* (Percoidei incertae sedis), new name and systematic position for Lates fajumensis Weiler, 1929, from the Eocene of the Fayum (Egypt)", *Neues Jahrbuch für Geologie und Paläontologie, Abhandlungen*, vol. 2, pp. 81–94, 1999.

[OTE 01a] OTERO O., "The oldest-known cyprinid fish of the Afro-Arabic Plate, and its paleobiogeographical implications", *Journal of Vertebrate Paleontology*, vol. 21, no. 2, pp. 386–388, 2001.

[OTE 01b] OTERO O., GAYET M., "Palaeoichthyofaunas from the Lower Oligocene and Miocene of the Arabian Plate: palaeoecological and palaeobiogeographical implications", *Palaeogeography, Palaeoclimatology, Palaeoecology*, vol. 165, no. 1, pp. 141–169, 2001.

[OTE 04] OTERO O., "Anatomy, systematics and phylogeny of both Recent and fossil latid fishes (Teleostei, Perciformes, Latidae)", *Zoological Journal of the Linnean Society*, vol. 141, pp. 81–133, 2004.

[OTE 08] OTERO O., VALETIN X., GARCIA G., "Cretaceous characiform fishes (Teleostei: Ostariophysi) from Northern Tethys: description of new material from the Maastrichtian of Provence (Southern France) and palaeobiogeographical implications", in CAVIN L., LONGBOTTOM A., RICHTER M. (eds), *Fishes and the Break-up of Pangaea*, Geological Society, London, 2008.

[OTE 09] OTERO O., PINTON A., TAISSO MACKAYE H. *et al.*, "Fishes and palaeogeography of the African drainage basins: relationships between Chad and neighbouring basins throughout the Mio-Pliocene", *Palaeogeography, Palaeoclimatology, Palaeoecology*, vol. 274, pp. 134–139, 2009.

[OTE 10] OTERO O., "What controls the freshwater fish fossil record? A focus on the Late Cretaceous and Tertiary of Afro-Arabia", *Cybium*, vol. 34, no. 1, pp. 93–113, 2010.

[OTE 11] OTERO O., "Current knowledge and new assumptions on the evolutionary history of the African lungfish, *Protopterus*, based on a review of its fossil record", *Cybium*, vol. 34, no. 1, pp. 93–113, 2011.

[OTE 15] OTERO O., PINTON A., CAPPETTA H. *et al.*, "A fish assemblage from the Middle Eocene from Libya (Dur At-Talah) and the Earliest Record of Modern African Fish Genera", *PLoS One*, vol. 10, p. e0144358, 2015.

[PAR 10] PARDO J.D., HUTTENLOCKER A.K., SMALL B.J. *et al.*, "The cranial morphology of a new genus of lungfish (Osteichthyes: Dipnoi) from the upper Jurassic Morrison Formation of North America", *Journal of Vertebrate Paleontology*, vol. 30, no. 5, pp. 1352–1359, 2010.

[PAN 13] PAN Y., SHA J., ZHOU Z. *et al.*, "The Jehol Biota: definition and distribution of exceptionally preserved relics of a continental Early Cretaceous ecosystem", *Cretaceous Research*, vol. 44, pp. 30–38, 2013.

[PAN 15] PAN Y., FÜRSICH F.T., ZHANG J. *et al.*, "Biostratinomic analysis of *Lycoptera* beds from the Early Cretaceous Yixian Formation, western Liaoning, China", *Palaeontology*, vol. 58, pp. 537–561, 2015.

[PAR 14] PARDO J.D., HUTTENLOCKER A.K., SMALL B.J., "An exceptionally preserved transitional lungfish from the Lower Permian of Nebraska, USA, and the origin of modern lungfishes", *PLoS ONE*, vol. 9, p. e108542, 2014.

[PAT 73] PATTERSON C., "Interrelationships of holosteans", in GREENWOOD P.H., MILES R.S., PATTERSON C. (eds), *Interrelationships of Fishes*, Linnean Society of London by Academic Press, 1973.

[PAT 75] PATTERSON C., "The distribution of mesozoic freshwater fishes, Biogéographie et liaisons intercontinentales au cours du Mésozoïque", *Mémoire du Muséum d'Histoire Naturelle*, Paris no. spec. A, Zool., 1975.

[PAT 81] PATTERSON C., "Significance of fossils in determining evolutionary relationships", *Annual Review of Ecology and Systematics*, vol. 12, pp. 195–223, 1981.

[PAT 84] PATTERSON C., "*Chanoides*, a marine eocene otophysan fish (Teleostei: Ostariophysi)", *Journal of Vertebrate Paleontology*, vol. 4, pp. 430–456, 1984.

[PAT 89] PATTERSON C., LONGBOTTOM, A.E., "An eocene Amiid fish from Mali, West Africa", *Copeia*, vol. 1989, no. 4, pp. 827–836, 1989.

[PEN 07] PENG Z., LUDWIG A., WANG D. *et al.*, "Age and biogeography of major clades in sturgeons and paddlefishes (Pisces: Acipenseriformes)", *Molecular Phylogenetics and Evolution*, vol. 42, pp. 854–862, 2007.

[PEN 15] PENG C., MURRAY A.M., BRINKMAN D.B. *et al.*, "A new species of *Sinamia* (Amiiformes, Sinamiidae) from the Early Cretaceous of Lanzhou Basin, Gansu, China", *Journal of Vertebrate Paleontology*, vol. 35, p. e902847, 2015.

[PER 10] PEREZ P.A., MALABARBA M.C., DEL PAPA C., "A new genus and species of Heroini (Perciformes: Cichlidae) from the early Eocene of southern South America", *Neotropical Ichthyology*, vol. 8, pp. 631–642, 2010.

[PLA 01] PLAZIAT J.-C., CAVAGNETTO C., KOENIGUER J.-C. *et al.*, "History and biogeography of the mangrove ecosystem, based on a critical reassessment of the paleontological record", *Wetlands Ecology and Management*, vol. 9, pp. 161–180, 2001.

[POH 15] POHL M., MILVERTZ F.C., MEYER A. *et al.*, "Multigene phylogeny of cyprinodontiform fishes suggests continental radiations and a rogue taxon position of *Pantanodon*", *Vertebrate Zoology*, vol. 65, no. 1, pp. 37–44, 2015.

[POY 95] POYATO-ARIZA F.J., "*Ichthyemidion*, a new genus for the elopiform fish "*Anaethalion*" *vidali*, from the Early Cretaceous of Spain: phylogenetic comments", *Comptes Rendus de l'Académie des Sciences Paris*, vol. 320, pp. 133–139, 1995.

[POY 98] POYATO-ARIZA F.J., TALBOT M.R., FREGENAL-MARTÍNEZ M.A.A. *et al.*, "First isotopic and multidisciplinary evidence for nonmarine coelacanths and pycnodontiform fishes: palaeoenvironmental implications", *Palaeogeography, Palaeoclimatology, Palaeoecology*, vol. 144, pp. 65–84, 1998.

[POY 02] POYATO-ARIZA F.J., WENZ S., "A new insight into pycnodontiform fishes", *Geodiversitas*, vol. 24, pp. 139–248, 2002.

[POY 04] POYATO-ARIZA F.J., WENZ S., "The new pycnodontid fish genus Turbomesodon, and a revision of Macromesodon based on new material from the Lower Cretaceous of Las Hoyas, Cuenca, Spain", in ARRATIA G., TINTORI A. (eds), *Mesozoic Fishes 3 – Systematics, Paleoenvironments and Biodiversity*, Dr. Friedrich Pfeil, Munich, 2004.

[POY 10] POYATO-ARIZA F.J., GRANDE T., DIOGO R., "Gonorynchiform interrelationships: historic overview, analysis, and revised systematics of the group", in GRANDE T., POYATO-ARIZA F.J., DIOGO R. (eds), *Gonorynchiformes and Ostariophysan Relationships: A Comprehensive Review*, Science Publishers, Enfield, 2010.

[POY 15] POYATO-ARIZA F.J., "Studies on pycnodont fishes (I): evaluation of their phylogenetic position among actinopterygians", *Revista Italiana di Paleontologica e Stratigrafia*, vol. 121, no. 3, pp. 329–343, 2015.

[PRI 99] PRICE G.D., "The evidence and implications of polar ice during the Mesozoic", *Earth-Science Reviews*, vol. 48, pp. 183–210, 1999.

[PRI 14] PRIEM F., "Sur des poissons fossiles et en particulier des siluridés du tertiaire supérieur et des couches récentes d'Afrique: Suivi de: Sur des poissons fossiles des terrains tertiaires d'eau douce et d'eau saumâtre de France et de Suisse", *Société géologique de France*, vol. 21, pp. 1–13, 1914.

[PUC 03] PUCÉAT E., LÉCUYER C., SHEPPARD S.M.F. *et al.*, "Thermal evolution of cretaceous Tethyan marine waters inferred from oxygen isotope composition of fish tooth enamels", *Paleoceanography*, vol. 18, pp. 1–12, 2003.

[QIA 09] QIAO T., ZHU M., "A new tooth-plated lungfish from the Middle Devonian of Yunnan, China, and its phylogenetic relationships", *Acta Zoologica*, vol. 90, pp. 236–252, 2009.

[RAT 14] RATMUANGKHWANG S., MUSIKASINTHORN P., KUMAZAWA Y., "Molecular phylogeny and biogeography of air sac catfishes of the Heteropneustes fossilis species complex (Siluriformes: Heteropneustidae)", *Molecular Phylogenetics and Evolution*, vol. 79, pp. 82–91, 2014.

[RAU 09] RAUHUT O.W.M., LÓPEZ-ARBARELLO A., "Considerations on the age of the Tiouaren Formation (Iullemmeden Basin, Niger, Africa), implications for Gondwanan Mesozoic terrestrial vertebrate faunas", *Palaeogeography, Palaeoclimatology, Palaeoecology*, vol. 271, pp. 259–267, 2009.

[REG 22] REGAN C.T., "The distribution of the fishes of the order Ostariophysi", *Bijdragen tot de Dierkunde*, vol. 22, pp. 203–208, 1922.

[ROC 03] ROCHA L.A., "Patterns of distribution and processes of speciation in Brazilian reef fishes", *Journal of Biogeography*, vol. 30, pp. 1161–1171, 2003.

[ROE 91] ROE L.J., "Phylogenetic and ecological significance of Channidae (Osteichthyes Teleostei) from the Early Eocene Kuldana Formation of Kohat, Pakistan", *Contributions from the Museum of Paleontology, University of Michigan*, vol. 28, pp. 93–100, 1991.

[ROM 16] ROMANO C., KOOT M.B., KOGAN I. *et al.*, "Permian–Triassic Osteichthyes (bony fishes): diversity dynamics and body size evolution", *Biological Reviews*, vol. 91, pp. 106–147, 2016.

[ROM 54] ROMER A.S., OLSON E.C., "Aestivation in a Permian lungfish", *Breviora*, vol. 30, pp. 1–8, 1954.

[ROS 74] ROSEN D.E., "Phylogeny and zoogeography of salmoniform fishes and relationships of *Lepidogalaxias salamandroides*", *American Museum of Natural History*, vol. 153, no. 2, pp. 265–326, 1954.

[ROS 81] ROSEN D.E., FOREY P.L., GARDINER B.G. *et al.*, "Lungfishes, Tetrapods, Paleontology and Plesiomorphy", *Bulletin of the American Museum of Natural History*, vol. 167, pp. 163–275, 1981.

[ROX 14] ROXO F.F., ALBERT J.S., SILVA G.S. *et al.*, "Molecular phylogeny and biogeographic history of the armored Neotropical catfish subfamilies Hypoptopomatinae, Neoplecostominae and Otothyrinae (Siluriformes: Loricariidae)", *PLoS One*, vol. 9, p. e105564, 2014.

[SAI 11] SAITOH K., SADO T., DOOSEY M.H. *et al.*, "Evidence from mitochondrial genomics supports the lower Mesozoic of South Asia as the time and place of basal divergence of cypriniform fishes (Actinopterygii: Ostariophysi)", *Zoological Journal of the Linnean Society*, vol. 161, pp. 633–662, 2011.

[SAN 09] SANTINI F., HARMON L.J., CARNEVALE G. *et al.*, "Did genome duplication drive the origin of teleosts? A comparative study of diversification in ray-finned fishes", *Evolutionary Biology*, vol. 9, pp. 1–15, 2009.

[SAU 80] SAUVAGE H.E., "Sur un *Prolebias* (*Prolebias Davidi*) des terrains tertiaires du nord de la Chine", *Bulletin de la Société Géologique de France*, vol. 3, no. 8, pp. 452–454, 1880.

[SCH 84a] SCHAAL S., "Oberkretazische Osteichthyes (Knochenfische) aus dem Bereich von Bahariya und Kharga, Ägypten, und ihre Aussagen zur Palökologie und Stratigraphie", *Berliner Geowissenschaftliche Abhandlungen, Reihe A: Geologie und Paläontologie*, vol. 53, pp. 1–79, 1984.

[SCH 52] SCHAEFFER B., "The evidence of the fresh-water fishes", *Bulletin of the American Museum of Natural History*, vol. 99, pp. 227–234, 1952.

[SCH 56] SCHAEFFER B., "Evolution in the subholostean fishes", *Evolution*, vol. 10, pp. 201–212, 1956.

[SCH 67] SCHAEFFER B., "Late Triassic fishes from the western United States", *Bulletin of the American Museum of Natural History*, vol. 135, no. 6, pp. 285–342, 1967.

[SCH 72] SCHAEFFER B., "A Jurassic fish from Antarctica", *American Museum Novitates*, vol. 2495, pp. 1–17, 1972.

[SCH 84b] SCHAEFFER B., "On the relationships of the Triassic-Liassic redfieldiiform fishes", *American Museum Novitates*, vol. 2795, pp. 1–18, 1984.

[SCH 16] SCHOCH R.R., SEEGIS D., "A Middle Triassic palaeontological gold mine: the vertebrate deposits of Vellberg (Germany)", *Palaeogeography, Palaeoclimatology, Palaeoecology*, vol. 459, pp. 249–267, 2016.

[SCH 81] SCHULTZE H.-P., "Das Schädeldach eines ceratodontiden Lungenfisches aus der Trias Süddeutschlands (Dipnoi, Pisces)", *Stuttgarter Beiträge zur Naturkunde, Serie B (Geologie und Paläontologie)*, vol. 70, pp. 1–31, 1981.

[SCH 91] SCHULTZE H.-P., "Lungfish from the El Molino (Late Cretaceous) and Santa Lucia formations in South-central Bolivia", in SUÁREZ-SORUCO R., *Fósiles y Facies de Bolivia 1 – Vertebrados*, Revista Técnica YPFB, La Paz, 1991.

[SCH 04] SCHULTZE H.-P., "Mesozoic sarcopterygians", in ARRATIA G., TINTORI A. (eds), *Mesozoic Fishes 3 – Systematics, Paleoenvironments and Biodiverity*, Dr. Friedrich Pfeil, Munich, 2004.

[SCH 09] SCHWIMMER D.R., "Giant coelacanths as the missing planktivores in Southeastern Late Cretaceous costal seas", *Geological Society of America Abstracts with Programs*, vol. 41, pp. 6, 2009.

[SCO 14] SCOTESE C.R., *Atlas of Phanerozoic Rainfall (Mollweide Projection), Volumes 1–6: PALEOMAP Project Atlas for ArcGIS*, PALEOMAP Project, Evanston, 2014.

[SCO 15] SCOTESE C.R. *Phanerozoic Temperature Curve*, PALEOMAP Project, Evanston, 2015.

[SEP 78] SEPKOSKI J.J., "A kinetic model of Phanerozoic taxonomic diversity I. Analysis of marine orders", *Paleobiology*, vol. 4, pp. 223–251, 1978.

[SEP 79] SEPKOSKI JR. J.J., "A kinetic model of Phanerozoic taxonomic diversity II. Early Phanerozoic families and multiple equilibria", *Paleobiology*, vol. 5, no. 3, pp. 222–251, 1979.

[SER 96] SERENO P.C., DUTHEI D.B., IAROCHENE M. *et al.*, "Predatory dinosaurs from the Sahara and late Cretaceous faunal differentiation", *Science*, vol. 272, pp. 986–991, 1996.

[SFE 15] SFERCO E., LÓPEZ-ARBARELLO A., MARIA BAEZ A., "Anatomical description and taxonomy of †*Luisiella feruglioi* (Bordas), new combination, a freshwater teleost (Actinopterygii, Teleostei) from the Upper Jurassic of Patagonia", *Journal of Vertebrate Paleontology*, vol. 35, p. e924958, 2015.

[SHE 92] SHEEHAN P.M., FASTOVSKY D.E., "Major extinctions of land-dwelling vertebrates at the Cretaceous-Tertiary boundary, eastern Montana", *Geology*, vol. 20, pp. 556–560, 1992.

[SHE 93] SHEEHAN P.M., FASTOVSKY D.E., "Major extinctions of land-dwelling vertebrates at the Cretaceous-Tertiary boundary, eastern Montana: comment and reply", *Geology*, vol. 21, pp. 92–93, 1993.

[SIG 97] SIGE B., BUSCALIONI A.D., DUFFAUD S. *et al.*, "Etat des données sur le gisement crétacé supérieur continental de Champ-Garimond (Gard, Sud de la France)", *Münchner Geowissenschaftliche Abhandlungen (A)*, vol. 34, pp. 111–130, 1997.

[SKR 14] SKRZYCKA R., "Revision of two relic actinopterygians from the Middle or Upper Jurassic Karabastau Formation, Karatau Range, Kazakhstan", *Alcheringa: An Australasian Journal of Palaeontology*, vol. 38, pp. 364–390, 2014.

[SMI 81] SMITH G.R., "Late Cenozoic freshwater fishes of North America", *Annual Review of Ecology and Systematics*, vol. 12, pp. 163–193, 1981.

[SMI 06] SMITH J.B., GRANDSTAFF B.S., ABDEL-GHANI M.S., "Microstructure of polypterid scales (Osteichthyes: Actinopterygii: Polypteridae) from the Upper Cretaceous Bahariya Formation, Bahariya oasis, Egypt", *Journal of Paleontology*, vol. 80, pp. 1179–1185, 2006.

[SMI 13] SMITH G.R., STEWART J.D., CARPENTER N.E., "Fossil and recent Mountain Suckers, Pantosteus, and significance of introgression in catostomin fishes of western United States", *Occasional papers of the Museum of Zoology*, University of Michigan, vol. 724, pp. 1–59, 2013.

[SOT 10] SOTO M., PEREA D., "Late Jurassic lungfishes (Dipnoi) from Uruguay, with comments on the systematics of Gondwanan ceratodontiforms", *Journal of Vertebrate Paleontology*, vol. 30, no. 4, pp. 1049–1058, 2010.

[SPA 04] SPARKS J.S., SMITH W., "Phylogeny and biogeography of cichlid fishes (Teleostei: Perciformes: Cichlidae)", *Cladistics*, vol. 20, pp. 501–517, 2004.

[STE 16a] STEARLEY R.F., SMITH G.R., "Fishes from the Mio-Pliocene Western Snake River plain and vicinity. I. Salmonid fishes from Mio-Pliocene lake sediments in the Western Snake River Plain and the Great Basin", *Miscellaneous Publications, Museum of Zoology, University of Michigan*, vol. 204, no. 1, pp. 1–43, 2016.

[STE 16b] STEVENS W.N., CLAESON K.M., STEVENS N.J., "Alestid (Characiformes: Alestidae) Fishes from the Late Oligocene Nsungwe Formation, Rukwa Rift Basin, of Tanzania", *Journal of Vertebrate Paleontology*, vol. 36, no 5, p. e1180299, 2016.

[STE 01] STEWART K.M., "The freshwater fish of Neogene Africa (Miocene–Pleistocene): systematics and biogeography", *Fish and Fisheries*, vol. 2, pp. 177–230, 2001.

[STR 25] STROMER E., "Der Rückgang der Ganoidfische von der Kreidezeit an", *Zeitschrift der Deutsche Geologische Gesellschafte*, vol. 77, pp. 348–371, 1925.

[STR 36] STROMER E., "Ergebnisse der Forschungsreisen Prof. E. Stromers in den Wüsten Ägyptens. VII Baharîje-Kessel und -Stufe mit deren Fauna und Flora. Eine ergänzende Zusammenfassung. *Abhandlungen der Bayerischen Akademie der Wissenschaften*", *Mathematisch-naturwissenschaftliche Abteilung Neue Folge*, vol. 33, pp. 1–102, 1936.

[STR 05] STROMER E., "Die Fischreste des mittleren und oberen Eocäns von Ägypten", *Beiträge zur Paläontologie und Geologie Österreichs-Ungarns und des Orients*, vol. 18, pp. 163–192, 1905.

[SU 85] SU D., "On Late Mesozoic fish fauna from Xinjiang (Sinkiang), China", *Institute of Vertebrate Paleontology and Paleoanthropology Memoirs*, vol. 17, pp. 61–136, 1985.

[SUL 02] SULLIVAN J.P., LAVOUÉ S., HOPKINS C.D., "Discovery and phylogenetic analysis of a riverine species flock of African electric fishes (Mormyridae: Teleostei)", *Evolution*, vol. 56, no. 3, pp. 597–616, 2002.

[SUL 06] SULLIVAN J.P., LUNDBERG J.G., HARDMAN M., "A phylogenetic analysis of the major groups of catfishes (Teleostei: Siluriformes) using rag1 and rag2 nuclear gene sequences", *Molecular Phylogenetics and Evolution*, vol. 41, no. 3, pp. 636–662, 2006.

[SUL 13] SULLIVAN J.P., MURIEL-CUNHA J., LUNDBERG J.G., "Phylogenetic relationships and molecular dating of the major groups of catfishes of the Neotropical superfamily Pimelodoidea (Teleostei, Siluriformes)", *Proceedings of the Academy of Natural Sciences of Philadelphia*, vol. 162, pp. 89–110, 2013.

[SUN 12] SUN Z., LOMBARDO C., TINTORI A. *et al.*, "*Fuyuanperleidus dengi* Geng *et al.*, 2012 (Osteichthyes, Actinopterygii) from the Middle Triassic of Yunnan province, South China", *Rivista Italiana di Paleontologia e Stratigrafia (Research In Paleontology and Stratigraphy)*, vol. 118, pp. 359–373, 2012.

[SUR 08] SURLYK F., MILÀN J., NOE-NYGAARD N., "Dinosaur tracks and possible lungfish aestivation burrows in a shallow coastal lake; lowermost Cretaceous, Bornholm, Denmark", *Palaeogeography, Palaeoclimatology, Palaeoecology*, vol. 267, pp. 292–304, 2008.

[SYT 99] SYTCHEVSKAYA E.K., "Freshwater fish fauna from the Triassic of Northern Asia", *Mesozoic Fishes*, vol. 2, pp. 445–468, 1999.

[SZA 16] SZABÓ M., GULYÁS P., ŐSI A., "Late Cretaceous (Santonian) *Atractosteus* (Actinopterygii, Lepisosteidae) remains from Hungary (Iharkút, Bakony Mountains)", *Cretaceous Research*, vol. 60, pp. 239–252, 2016.

[TAB 63] TABASTE N., "Etude de restes de poissons du Crétacé saharien", *Mémoire IFAN, Mélanges Ichthyologiques*, vol. 68, pp. 437–485, 1963.

[TAN 11] TANG K.L., AGNEW M.K., CHEN W.-J. *et al.*, "Phylogeny of the gudgeons (Teleostei: Cyprinidae: Gobioninae)", *Molecular Phylogenetics and Evolution*, vol. 61, pp. 103–124, 2011.

[TAV 84] TAVERNE L., "À propos de *Chanopsis lombardi* du Crétacé inférieur du Zaïre (Teleostei, Osteoglossiformes)", *Revue de zoologie africaine*, vol. 98, pp. 578–590, 1984.

[TAV 95] TAVERNE L., "Description de l'appareil de Weber du téléostéen marin *Clupavus maroccanus* et ses implications phylogénétiques", *Belgian Journal Of Zoology*, vol. 125, pp. 267–282, 1995.

[TAV 98] TAVERNE L., "Les Ostéoglossomorphes marins de l'Eocène du Monte Bolca (Italy): *Monopterus* Volta 1796, *Thrissopterus* Heckel, 1856 et *Foreyichthys* Taverne, 1979. Considérations sur la phylogénie des téléostéens Ostéoglossomorphes", in *Studi e Ricerche sui Giacimenti Terziari di Bolca. VII*, Museo Civico di Storia Naturale, Verona, 1998.

[TAV 01] TAVERNE L., "Position systématique et relations phylogénétiques de *Paraclupavus* ('*Leptolepis*') *caheni*, téléostéen marin du Jurassique moyen de Kisangani (Calcaires de Songa, étage de Stanleyville), République Démocratique du Congo. Musée royal de l'Afrique centrale, Tervuren (Belg.)", *Rapport Annuel 1999–2000, Département du Géologie et Mineralogie*, pp. 55–76, 2001.

[TAV 03] TAVERNE L., FILLEUL A., "Osteology and relationships of the genius *Spaniodon* (Teleostei, Salmoniformes) from the Santonian (Upper Cretaceous) of Lebanon", *Palaeontology*, vol. 46, pp. 927–944, 2003.

[TAV 05] TAVERNE L., "Les poissons crétacés de Nardo. 20. *Chanoides chardoni* sp. nov. (Teleostei, Ostariophysi, Otophysi)", *Bolletino del Museo Civico di Storia Naturale di Verona*, vol. 29, pp. 39–54, 2005.

[TAV 07] TAVERNE L., NOLF D., FOLIE A., "On the presence of the osteoglossid fish genus *Scleropages* (Teleostei, Osteoglossiformes) in the continental Paleocene of Hainin (Mons Basin, Belgium)", *Belgian Journal of zoology*, vol. 137, pp. 89, 2007.

[TAV 09a] TAVERNE L., KUMAR K., RANA R.S., "Complement to the study of the Indian Paleocene osteoglossid fish genus *Taverneichthys* (Teleostei, Osteoglossomorpha)", *Bulletin de l'Institut Royal des Sciences Naturelles de Belgique, Sciences de la Terre*, vol. 79, pp. 155–160, 2009.

[TAV 09b] TAVERNE L., "On the presence of the osteoglossid *Scleropages* in the Paleocene of Niger, Africa (Teleostei, Osteoglossomorpha)", *Bulletin de l'Institut royal des Sciences naturelles de Belgique, Sciences de la Terre*, vol. 79, pp. 161–167, 2009.

[TAV 11a] TAVERNE L., "Ostéologie et relations phylogénétiques de *Steurbautichthys* ("*Pholidophophorus*") *aequatorialis* gen. nov. (Teleostei, "Pholidophoriformes") du Jurassique moyen de Kisangani, en République Démocratique du Congo", *Bulletin de L'Institut Royal des Sciences Naturelles de Belgique, Sciences de la Terre*, vol. 81, pp. 129–173, 2011.

[TAV 11b] TAVERNE L., "Ostéologie et relations de *Catervariolus* (Teleostei, "Pholidophoriformes") du Jurassique moyen de Kisangani (Formation de Stanleyville) en République Démocratique du Congo", *Bulletin de l'Institut Royal des Sciences Naturelles de Belgique, Sciences de la Terre*, vol. 81, pp. 175–212, 2011.

[TAV 11c] TAVERNE L., "Ostéologie et relations de *Ligulella* (Halecostomi, Ligulelliformes nov. ord.) du Jurassique moyen de Kisangani (Formation de Stanleyville) en République Démocratique du Congo", *Bulletin de l'Institut Royal des Sciences Naturelles de Belgique, Sciences de la Terre*, vol. 81, pp. 213–233, 2011.

[TAV 12] TAVERNE L., CAPASSO, L., "Osteology and relationships of *Prognathoglossum kalassyi* gen. and sp. nov. (Teleostei, Osteoglossiformes, Pantodontidae) from the marine Cenomanian (Upper Cretaceous) of En Nammoura (Lebanon)", *Cybium, International Journal of Ichthyology*, vol. 36, pp. 563–575, 2012.

[TAV 13] TAVERNE L., "Osteology and relationships of *Songaichthys luctacki* gen. and sp. nov. (Teleostei, Ankylophoriformes ord. nov.) from the Middle Jurassic (Songa Limestones) of Kisangani (Democratic Republic of Congo)", *Geo-Eco-Trop*, vol. 37, no. 1, pp. 33–52, 2013.

[TAV 14a] TAVERNE L., "Ostéologie et position systématique de *Songanella callida* (Teleostei, Catervarioliformes nov. ord.) du Jurassique moyen de Kisangani (Formation de Stanleyville, Calcaires de Songa) en République Démocratique du Congo", Geo-*Eco-Trop*, vol. 37, no. 1, pp. 1–32, 2014.

[TAV 14b] TAVERNE L., "Osteology and relationships of *Kisanganichthys casieri* gen. and sp. nov. (Teleostei, Catervariolidae) from the Middle Jurassic (Stanleyville Formation) of Kisangani (Congo R.D.) Comments on the systematic position of Catervarioliformes", *Geo-Eco-Trop*, vol. 38, no. 2, pp. 241–258, 2014.

[TAV 14c] TAVERNE L., "Osteology and phylogenetic relationships of *Congophiopsis lepersonnei* gen. nov. (Halecomorphi, Ionoscopiformes) from the Songa Limestones (Middle Jurassic, Stanleyville Formation), Democratic Republic of Congo", *Geo-Eco-Trop*, vol. 38, no. 2, pp. 223–240, 2014.

[TRA 10] TRAQUAIR R.H., "Les poissons wealdiens de Bernissart", *Mémoire du Musée Royal d'Histoire Naturelle de Belgique*, vol. 6, pp. 1–65, 1910.

[TUR 99] TURGEON J., ESTOUP A., BERNATCHEZ L., "Species flock in the North American Great Lakes: molecular ecology of lake Nipigon ciscoes (Teleostei: Coregonidae: *Coregonus*)", *Evolution*, vol. 4, pp. 1857–1871, 1999.

[TUR 03] TURGEON J., BERNATCHEZ L., "Reticulate evolution and phenotypic diversity in North American ciscoes, *Coregonus* ssp. (Teleostei: Salmonidae): implications for the conservation of an evolutionary legacy", *Conservation Genetics*, vol. 4, no. 1, pp. 67–81, 2003.

[TUR 16] TURNER S., LONG J., "The Woodward factor: Arthur Smith Woodward's legacy to geology in Australia and Antarctica", *Geological Society, London, Special Publications*, vol. 430, pp. 261–288, 2016.

[UNM 01] UNMACK P.J., "Biogeography of Australian freshwater fishes", *Journal of biogeography*, vol. 28, pp. 1053–1089, 2001.

[UNM 10] UNMACK P.J., DOWLING T.E., "Biogeography of the genus *Craterocephalus* (Teleostei: Atherinidae) in Australia", *Molecular Phylogenetics and Evolution*, vol. 55, no. 3, pp. 968–984, 2010.

[VAV 14] VAVREK M.J., MURRAY A.M., BELL P.R., "An early late cretaceous (Cenomanian) sturgeon (Acipenseriformes) from the Dunvegan formation, northwestern Alberta, Canada", *Canadian Journal of Earth Sciences*, vol. 51, pp. 677–681, 2014.

[VEG 12] VEGA G.C., WIENS J.J., "Why are there so few fish in the sea?", *Proceedings of the Royal Society of London B: Biological Sciences*, vol. 279, pp. 2323–2329, 2012.

[VÉR 15] VÉRARD C., HOCHARD C., BAUMGARTNER P.O. *et al.*, "3D palaeogeographic reconstructions of the Phanerozoic versus sea-level and Sr-ratio variations", *Journal of Palaeogeography*, vol. 4, no. 1, pp. 64–84, 2015.

[VIC 96] VICKERS-RICH P., MOLNAR R.E., "The foot of a bird from the Eocene Redbank Plains Formation of Queensland, Australia", *Alcheringa*, vol. 20, no. 1, pp. 21–29, 1996.

[VUL 08] VULLO R., NÉRAUDEAU D., "Cenomanian vertebrate assemblages from southwestern France: a new insight into the European mid-Cretaceous continental fauna", *Cretaceous Research*, vol. 29, pp. 930–935, 2008.

[VUL 16] VULLO R., ALLAIN R., CAVIN L., "Convergent evolution of jaws between spinosaurid dinosaurs and pike conger eels", *Acta Palaeontologia Polonica*, vol. 61, no. 4, pp. 825–828, 2016.

[WAD 08] WADDELL L.M., MOORE T.C., "Salinity of the Eocene Arctic Ocean from oxygen isotope analysis of fish bone carbonate", *Paleoceanography*, vol. 23, pp. 1–14, 2008.

[WAL 71] WALDMAN M., "Fish from the freshwater Lower Cretaceous of Victoria, Australia with comments on the palaeo-environment", *Special Paper in Palaeontology*, vol. 9, pp. 1–62, 1971.

[WEI 35] WEILER W., "Ergebnisse der Forschungsreisen Prof. E. Stromers in den Wüsten Ägyptens. II. Wirbeltierreste der Baharîje-Stufe (unterstes Cenoman). Neue Untersuchungen an den Fishresten. Abhandlungen der Bayerischen Akademie der Wissenschaften", *Mathematisch-naturwissenschaftliche Abteilung Neue Folge*, vol. 32, pp. 1–57, 1935.

[WEN 75] WENZ S., "Un nouveau coelacanthidé du Crétacé inférieur du Niger, remarques sur la fusion des os dermiques", *Colloque International CNRS: Problèmes Actuels de Paléontologie-Evolution des Vertébrés*, Paris, 1975.

[WEN 92] WENZ S., BRITO P.M., "Première découverte de Lepisosteidae (Pisces, Actinopterygii) dans le Crétacé inférieur de la Chapada do Araripe (N-E du Brésil). Conséquence sur la phylogénie des Ginglymodi", *Comptes Rendus de l'Académie des Sciences Paris*, vol. 314, pp. 1519–1525, 1992.

[WEN 96] WENZ S., BRITO P.M., "New data about the lepisosteids and semionotids from the Early Cretaceous of Chapada do Araripe (NE Brazil): phylogenetic implications", in ARRATIA G., SCHULTZE H.-P. (eds), *Mesozoic Fishes 1 – Systematics and Paleoecology*, Dr. Friedrich Pfeil, Munich, 1996.

[WEN 99] WENZ S., "*Pliodetes nigeriensis*, gen. nov. et sp. nov., a new semionotid fish from the Lower Cretaceous of Gadoufaoua (Niger Republic): phylogenetic comments", in ARRATIA G., SCHULTZE H.-P. (eds), *Mesozoic Fishes 2 – Systematics and Fossil Record*, Dr. Friedrich Pfeil, Munich, 1999.

[WER 94] WERNER C., "Die kontinentale Wirbeltierfauna aus der unteren Oberkreide des Sudan (Wadi Milk Formation)", *Berliner geowissenschaftliche Abhandlungen Reihe E*, vol. 13, pp. 221–249, 1994.

[WER 97] WERNER C., GAYET M., "New fossil Polypteridae from the Cenomanian of Sudan. An evidence of their high diversity in the Early Late Cretaceous", *Cybium*, vol. 21, pp. 67–81, 1997.

[WHI 36] WHITE E., "The name of a fossil cat-fish", *Geological Magazine, London*, vol. 74, pp. 319–325, 1936.

[WIL 77] WILSON M.V.H., *Middle Eocene freshwater fishes from British Columbia*. Royal Ontario Museum, Toronto, 1977.

[WIL 80] WILSON M.V.H., "Oldest known *Esox* (Pisces: Esocidae), part of a new Paleocene teleost fauna from western Canada", *Canadian Journal of Earth Science*, vol. 17, pp. 307–312, 1980.

[WIL 91] WILSON M.V.H., WILLIAMS R., "A new Paleocene smelt (Teleostei: Osmeridae) from the Paskapoo Formation of Alberta, and comments on osmerid phylogeny", *Journal of Vertebrate Paleontology*, vol. 11, pp. 434–451, 1991.

[WIL 92a] WILSON M.V.H., WILLIAMS R., *Phylogenetic, Biogeographic, and Ecological Significance of Early Fossil Records of North American Freshwater Teleostean Fishes. Systematics, Historical Ecology, and North American Freshwater Fishes*, Stanford University Press, Stanford, 1992.

[WIL 92b] WILSON M.V.H., BRINKMAN D.B., NEUMAN A.G., "Cretaceous Esocoidei (Teleostei): early radiation of the Pikes in North American fresh waters", *Journal of Paleontology*, vol. 66, pp. 839–846, 1992.

[WIL 96] WILSON M.V.H., "Taphonomy of a mass-death layer of fishes in the Paleocene Paskapoo Formation at Joffre Bridge, Alberta Canada", *Canadian Journal of Earth Sciences*, vol. 33, pp. 1487–1498, 1996.

[WIL 99] WILSON M.V.H., LI G.Q., "Osteology and systematic position of the Eocene salmonid *Eosalmo driftwoodensis* Wilson from the western North America", *Zoological Journal of the Linnean Society*, vol. 125, pp. 279–311, 1999.

[WIL 08] WILSON M.V.H., MURRAY A.M., "Osteoglossomorpha: phylogeny, biogeography, and fossil record and the significance of key African and Chinese fossil taxa", in CAVIN L., LONGBOTTOM A., RICHTER M. (eds), *Fishes and the Break-up of Pangaea*, Geological Society, London, 2008.

[WOO 08] WOODWARD A.S., "On some fossil fishes discovered by Prof. Ennes de Souza in the Cretaceous Formation at Ilhéos (State of Bahia), Brazil", *Quarterly Journal of the Geological Society of London*, vol. 64, pp. 358–362, 1908.

[WOO 16] WOODWARD A.S., "The fossil fishes of the English Wealden and Purbeck formations. Part 1", *Monograph of the Palaeontographical Society*, vol. 69, pp. 1–48, 1916.

[XU 09] XU G.H., CHANG M.M., "Redescription of †*Paralycoptera wui* Chang & Chou, 1977 (Teleostei: Osteoglossoidei) from the Early Cretaceous of eastern China", *Zoological Journal of the Linnean Society*, vol. 157, pp. 83–106, 2009.

[XU 11] XU G.-H., GAO K.-Q., "A new scanilepiform from the Lower Triassic of northern Gansu Province, China, and phylogenetic relationships of non-teleostean Actinopterygii", *Zoological Journal of the Linnean Society*, vol. 161, pp. 595–612, 2011.

[XU 14] XU G.-H., GAO K.-Q., FINARELLI J.A., "A revision of the Middle Triassic scanilepiform fish *Fukangichthys longidorsalis* from Xinjiang, China, with comments on the phylogeny of the Actinopteri", *Journal of Vertebrate Paleontology*, vol. 34, pp. 747–759, 2014.

[YAB 94] YABUMOTO Y., "Early Cretaceous Freshwater Fish Fauna in Kyushu, Japan", *Bulletin of the Kitakyushu Museum of Natural History and Human History A*, vol. 13, pp. 107–254, 1994.

[YAB 98] YABUMOTO Y., YANG S., KIM T., "Early Cretaceous freshwater fishes from Japan and Korea", *Journal of the Paleontological Society of Korea*, vol. 22, pp. 119, 1998.

[YAB 00] YABUMOTO Y., UYENO T., "*Inabaperca taniurai*, a new genus and species of Miocene percoid fish from Tottori Prefecture, Japan", *Bulletin of the National Science Museum. Series C, Geology & Paleontology 26*, pp. 93-106, 2000.

[YAB 05] YABUMOTO Y., "Early Cretaceous freshwater fishes from the Tetori Group, central Japan", *Bulletin of the Kitakyushu Museum of Natural History and Human History A*, vol. 3, pp. 135–143, 2005.

[YAB 08] YABUMOTO Y., "A new Early Cretaceous osteoglossomorph fish from Japan, with comments on the origin of the Osteoglossiformes2", *Mesozoic Fishes*, vol. 4, pp. 217–228, 2008.

[YAB 08] YABUMOTO Y., "A new Mesozoic coelacanth from Brazil (Sarcopterygii, Actinistia)", *Paleontological Research*, vol. 12, pp. 329–343, 2008.

[YAB 14] YABUMOTO Y., "*Sinamia kukurihime*, a new Early Cretaceous amiiform fish from Ishikawa, Japan", *Paleontological Research*, vol. 18, pp. 211–223, 2014.

[ZHA 06a] ZHANG J., "Morphology and phylogenetic relationships of †*Kuntulunia* (Teleostei: Osteoglossomorpha)", *Journal of Vertebrate Paleontology*, vol. 18, pp. 280–300, 2006.

[ZHA 06b] ZHANG J., "Phylogeny of osteoglossomorpha", *Vertebrata PalAsiatica*, vol. 44, pp. 43, 2006.

[ZHO 16] ZHOU C., WANG X., GAN X. *et al.*, "Diversification of Sisorid catfishes (Teleostei: Siluriformes) in relation to the orogeny of the Himalayan Plateau", *Science Bulletin*, vol. 61, pp. 991–1002, 2016.

Index

Topics, Families and Genera

Printed in the United States
By Bookmasters